CHAPMAN & HALL/CRC
Monographs and Surveys in
Pure and Applied Mathematics   102

INTRODUCTION TO

HAMILTONIAN FLUID

DYNAMICS

AND

STABILITY THEORY

GORDON E. SWATERS

CHAPMAN & HALL/CRC
Monographs and Surveys in
Pure and Applied Mathematics   102

INTRODUCTION TO

HAMILTONIAN FLUID

DYNAMICS

AND

STABILITY THEORY

GORDON E. SWATERS

CRC Press
Taylor & Francis Group
Boca Raton London New York

CRC Press is an imprint of the
Taylor & Francis Group, an **informa** business

A CHAPMAN & HALL BOOK

CRC Press
Taylor & Francis Group
6000 Broken Sound Parkway NW, Suite 300
Boca Raton, FL 33487-2742

First issued in paperback 2019

© 2000 by Taylor & Francis Group, LLC
CRC Press is an imprint of Taylor & Francis Group, an Informa business

No claim to original U.S. Government works

ISBN-13: 978-1-58488-023-3 (hbk)
ISBN-13: 978-0-367-39940-5 (pbk)

Visit the Taylor & Francis Web site at
http://www.taylorandfrancis.com

and the CRC Press Web site at
http://www.crcpress.com

To my son Sean G. Swaters.
Long may love reign over him.

Contents

About the author

Gordon E. Swaters is Professor of Mathematical Sciences and Director of the Applied Mathematics Institute at the UNIVERSITY OF ALBERTA. He is also an adjunct Professor in the Department of Earth & Atmospheric Sciences.

Gordon obtained the *B. Math. (Hons.)* degree from the UNIVERSITY OF WATERLOO and the *Ph.D.* from the UNIVERSITY OF BRITISH COLUMBIA.

Gordon's research is focussed on understanding nonlinear wave and eddy processes in planetary-scale atmosphere and ocean dynamics. In particular, he has made contributions to understanding the destabilization of buoyancy and density-driven ocean currents for which he was awarded the 1994 *President's Prize* of the Canadian Meteorological and Oceanographic Society (*CMOS*) - the most prestigious research award given by that society. Ten years earlier, in 1984, Gordon was awarded the *Graduate Student Prize* by *CMOS* for showing that the formation of large scale ocean eddies in the northeast Pacific ocean could be correlated to current changes forced by El Nino/Southern Oscillation events in the tropical Pacific.

In addition, Gordon has been a recipient of the *Faculty of Science Research Award, McCalla Research Professorship* and a *Killam Annual Professorship* at the UNIVERSITY OF ALBERTA.

Gordon was also a member of the Canadian Scientific Team of the Surface Velocity Drifter Program part of the *World Ocean Circulation Experiment*. This multinational research program made an important contribution in helping to understand the role of the oceans in Earth's evolving climate.

About the Authors

Gordon E. Swaters is a Professor in the Department of Mathematics and Director of the Applied Mathematics Institute at the University of Alberta. He is also an adjunct Professor in the Department of Earth and Atmospheric Sciences.

Gordon obtained the B.Math. (Hons) degree from the University of Waterloo and the Ph.D. from the University of British Columbia ...

[text illegible]

Acknowledgments

Science is not a competition between individuals to see who gets the biggest grants or publishes the most papers. Blair Kinsman's eloquent preface to his book *Water Waves* expressed the point that Science, in addition to its obvious utilitarian importance in human affairs, is about a community of men and women engaged in conversation. There are two members of this community that I would like to publicly acknowledge here. The first is Lawrence Mysak who taught me that the give and take of this conversation is not about who is better, but is simply the process of discovering the truth. The second is Paul LeBlond who taught me not to confuse the language of this conversation for the conversation itself. It is the spirit of community in Science that I cherish the most and I thank Lawrence and Paul for introducing me to it.

I would also like to thank the Natural Sciences and Engineering Research Council of Canada, the Department of Fisheries and Oceans of Canada, and the Atmospheric Environment Service of Canada for their financial support of my research.

Chapter 1

Introduction

The study of the dynamics of fluids remains one of the most difficult areas in all of science. From the generation of water waves seen at the beach, to the formation of the Great Red Spot on Jupiter, there are no completely accepted explanations. In some respects this may seem quite remarkable since, so far as we are concerned here, the governing equations of fluid dynamics, the *Navier-Stokes* equations have been known for well over a hundred and fifty years (see the remarks by Lamb, 1932 concerning the contributions of Navier, 1827; Poisson, 1831; de Saint Venant, 1843; Stokes, 1845). The Navier-Stokes equations are a system of nonlinear partial differential equations in which the nonlinear terms play an essential role in determining the evolution of the flow. The enormous technical and conceptual difficulties associated with understanding nonlinear dynamical systems may help to explain why fluid dynamics temporarily lost its appeal as a fashionable research area in the middle third of the 20*th* century.

The fact that fluid dynamics is once again an attractive research area is undoubtedly due to many reasons. Perhaps the most important is the introduction of efficient high resolution *numerical simulations* into fluid dynamics as a research tool. This is a tool that is nowhere near exhausting its usefulness. Of course, while computing power is necessary, it is not sufficient for a clear understanding of fluid dynamics. As in any area of inquiry, fluid dynamics can only proceed by the interplay of experience *and* theory.

Perhaps another reason why recent progress has been made in fluid dynamics research is that two new, seemingly quite contrary, concepts have been introduced. These are the notions of *chaos* and the *soliton*. They are contrary in the sense that chaos, understood as the sensitive dependence on initial conditions, implies the rapid decorrelation between the initial flow configuration and the flow configuration a short time later. The soliton, with its global nonlinear stability, is a robust finite-amplitude solution which never changes in time (modulo, of course, its translation).

It may well be that these two concepts are the key to understanding the evolution of a turbulent flow. It is probable that the formal notion of sensitive dependence to initial conditions is the mathematical reason why turbulent fluid flows evolve with respect to time in such a highly irregular way. On the other hand, numerical simulations of turbulent flows suggest that while the temporal evolution appears highly decorrelated, there appear to be relatively long-lived, as compared to typical eddy-turning time-scales, spatial coherent structures. These coherent structures often take the appearance of eddy-like flow features. It is possible to interpret the soliton as a very special example of a coherent structure. Thus a picture emerges of turbulent fluid flow which consists of sequence of rapidly evolving interactions of coherent wave and eddy-like structures. If this is an appropriate view of the onset and development of turbulence, then it would seem useful to understand and attempt to exploit the mathematical formalism upon which coherent structure and modern dynamical systems theory is constructed.

Modern soliton theory draws heavily on the Hamiltonian structure of the governing equations. For example, soliton stability theory (e.g., Benjamin, 1972; Bona, 1975) and soliton modulation theory (e.g., Kaup & Newell, 1978) directly exploit the Hamiltonian structure associated with the governing equations. Arnol'd's hydrodynamic stability theorems (Arnol'd, 1965, 1969; Holm et al., 1985) are also best understood as a consequence of the Hamiltonian structure of the governing equations.

In *finite dimensional mechanics*, nonlinear stability in the sense of Liapunov of an equilibrium solution to a conservative dynamical system is usually established by proving that the equilibrium solution corresponds to a local extremum of the energy (which is usually an invariant of the dynamics). This procedure has four key steps to it.

1. It is necessary to show that the equilibrium solution satisfies the first order necessary conditions for an energy extremum. That is, it is necessary to show that the first differential of the energy evaluated at the equilibrium solution is identically zero for all suitable perturbations.

2. One must show that the second differential of the energy evaluated at the equilibrium solution is negative or positive definite for all suitable perturbations. It is a classical result, apparently originating with Lagrange, that the second differential of an invariant which satisfies the first order necessary conditions is, when evaluated at the equilibrium solution, itself an invariant of the appropriate linear stability equations.

3. Since the second differential of the energy evaluated at the equilibrium solution is a quadratic form with respect to the perturbation field, it can be used to provide a norm on the perturbation field. Thus, if the definiteness of the second differential of the energy evaluated at the equilibrium solution can be proven, then it is usually straightforward to establish the linear stability in the sense of Liapunov for the equilibrium solution.

4. Because finite-dimensional vector spaces are compact, the definiteness of the second differential of the energy evaluated at the equilibrium solution implies

that the equilibrium corresponds to a local extremum of the energy. Thus the *nonlinear* stability in the sense of Liapunov of the equilibrium solution can be proven.

In fluid dynamics, it turns out that stationary flows do not in general satisfy the first order necessary conditions for an energy extremum. Thus the classical method breaks down and linear stability, assuming it exists for a given flow, has to be established with another argument. Another problem in fluid dynamics is that the appropriate function spaces are infinite dimensional so that compactness is lost. This means that even if a functional was found for which the stationary flow satisfied the appropriate first order necessary conditions *and* for which the second variation was definite, topological difficulties prevent immediately concluding that the functional is locally convex at the stationary flow. (In infinite dimensions it is necessary to work with *variations* rather than *differentials*.) Thus it could not be concluded that the stationary solution is a local extremum of the functional and, consequently, nonlinear stability could not be established.

Arnol'd's (1965, 1969) contribution was to construct an invariant pseudo energy functional which satisfied the first order necessary conditions for an arbitrary two dimensional steady flow of an incompressible fluid. Further, he established *sufficient conditions* on the equilibrium solution which would ensure the definiteness of the second variation of the pseudo energy evaluated at the steady flow. He went on to show that the conditions he found for establishing the *positive definiteness* of the pseudo energy reduced, in the limit of parallel shear flow, to the well-known *Fjortoft's stability theorem* (e.g., Drazin & Reid, 1981). Based on these sufficient conditions, Arnol'd was able to formally introduce appropriate convexity hypotheses on the density of the pseudo energy functional that could, in principle, establish nonlinear stability in the sense of Liapunov for a given steady flow.

From a pedagogical point of view, Arnol'd's stability method as it was originally presented suffered from one problem. The construction of the pseudo energy functional appeared to be based on an *ad hoc* procedure. It was not clear just how to generalize Arnol'd's approach for other conservative infinite dimensional dynamical systems. It is now known, however, that Arnol'd's approach directly exploits the Hamiltonian structure of the governing equations (e.g., Arnol'd, 1978; Holm *et al.*, 1985). From the point of view of wanting to apply Arnol'd's stability argument to other physical problems, it is best to present his approach as a consequence of the Hamiltonian structure of the governing equations.

In Chapter 2 a review of the symplectic Hamiltonian formulation of the nonlinear pendulum and stability characteristics of the equilibrium solutions is given. This finite dimensional example serves to introduce the notation, jargon and most of the necessary mathematical properties needed in the infinite dimensional Hamiltonian formalism to be developed in the subsequent chapters. It is shown how equilibrium solutions of an invertible finite dimensional Hamiltonian system satisfy the first-order necessary conditions for an extremal of the Hamiltonian. This variational principle is then exploited to derive sufficient convexity conditions on the Hamiltonian capable, in principle, of establishing the stability of an equilibrium solution to a general canonical finite dimensional Hamiltonian system. The general theory is applied to the nonlinear pendulum and the stability characteristics of the equilibrium solutions is discussed in the context of the Hamiltonian structure.

We develop the infinite dimensional Hamiltonian formulation of the two dimensional incompressible Euler equations in Chapter 3. The chapter begins with a derivation of the vorticity equation formulation of the two dimensional Euler equations and boundary conditions appropriate for simply and multiply connected domains. We then develop a general theory for infinite dimensional Hamiltonian systems of partial differential equations from an axiomatic viewpoint as motivated by the results obtained for the nonlinear pendulum. The Hamiltonian structure for the two dimensional vorticity equation is introduced and each required property of the Hamiltonian structure is rigorously established - including the Jacobi identity for the proposed Poisson bracket. Chapter 3 also includes a detailed derivation of the Poisson bracket (written in terms of Eulerian variables) for the vorticity equation as a systematic reduction of the canonical Poisson bracket (written in terms of Lagrangian variables) for the two dimensional Euler equations.

Noether's Theorem describes the connection between an invariance property of the Hamiltonian structure and a corresponding time invariant functional. A derivation of Noether's Theorem, in the context of scalar Hamiltonian systems, is presented in Chapter 3. We apply Noether's Theorem to the Hamiltonian structure of the two dimensional vorticity equation to show the connection between the invariance with respect to arbitrary linear spatial and temporal translations and rotational translations and, respectively, the conservation of linear momentum, energy and angular momentum. The invariant functionals associated with Noether's Theorem are very useful in deriving alternate variational principles for steady flows with special symmetries.

In Chapter 4 we turn to applying the Hamiltonian formulation for the two dimensional incompressible Euler equations to the question of the stability of steady solutions. We begin by characterizing steady solutions in terms of the relationship between the vorticity and stream function fields. It is shown how to construct a variational principle for arbitrary steady solutions based on the Hamiltonian structure. The general linear stability problem is described.

Much of classical hydrodynamic stability theory centres on looking for exponentially growing/decaying solutions to the Rayleigh stability equation. We give a brief derivation of this equation for parallel shear flows and briefly describe the classical stability theorems of Rayleigh and Fjortoft. A detailed linear and nonlinear stability analysis is given for arbitrary steady solutions based on the Hamiltonian structure. We show how Arnol'd's Stability Theorems reduce to Fjortoft's Stability Theorem.

It turns out that there is a coupling between the symmetry associated with the boundary conditions and the class of steady flows which can, in principle, satisfy the Arnol'd stability criteria. This result is expressed through *Andrews's Theorem* (Andrews, 1984). We discuss the implications of Andrews's Theorem in the context of two dimensional flow in a periodic channel and on the entire plane.

For some flows with special symmetries it is possible to exploit alternate variational principles to examine the stability properties of these flows. We discuss the cases corresponding to parallel shear flow and radially symmetric flows in this context. It is shown that parallel shear flows satisfy the first order necessary conditions for an extremal of a constrained linear impulse functional. We go on to show that the conditions which ensure that the second variation of this constrained impulse functional are definite (which in turn establishes the linear stability of the shear flow) are equivalent to Rayleigh's Inflection Point Theorem. The nonlinear generalizations of these results are derived.

We show that circular flows satisfy the first order conditions for an extremal of a constrained angular momentum functional. The conditions which ensure that the second variation of this constrained angular momentum functional are definite (which in turn establishes the stability of the circular flow) are equivalent to Rayleigh's Inflection Point Theorem applied to these flows. We establish the nonlinear generalization of these linear stability conditions.

One unsatisfactory feature of the two dimensional vorticity equation is that it is, of course, incapable of modelling any three dimensional effects such as vortex tube stretching/compression. The dilation and compression of vortex tubes and filaments is a fundamental characteristic of turbulent three dimensional fluid flows. Another feature that the two dimensional vorticity equation does not include, in the absence of a mean flow, is a term corresponding to a background vorticity gradient. The most important consequence of this property is that there are no nontrivial linear plane wave solutions to the two dimensional vorticity equation.

The simplest generalization of the two dimensional vorticity equation which includes these effects is the *Charney-Hasegawa-Mima (CHM)* equation. The *CHM* equation may be considered a canonical equation for quasi-two dimensional flow in the presence of a relatively constant background vorticity gradient. It is usually obtained via a formal asymptotic reduction of the relevant governing equations. For example, Hasegawa & Mima (1978) have shown that it describes the leading order evolution of the electrostatic potential for a weakly magnetized nonuniform plasma. In the context of geophysical fluid dynamics, the *CHM* equation describes the low wavenumber and frequency evolution of the leading order dynamic pressure for a rapidly rotating fluid of finite thickness (Charney, 1947; see also LeBlond & Mysak, 1978 or Pedlosky, 1987). The *CHM* equation possesses linear dispersive wave solutions known as *Rossby* waves and steadily-travelling isolated dipole vortex solutions known as *modons*.

In Chapter 5 we give a derivation of the *CHM* equation based on the shallow water equations for a differential rotating inviscid fluid. The Hamiltonian structure is introduced and is exploited to give variational principles for steady and steadily-travelling solutions. The normal mode stability problem is examined for parallel shear flow solutions to the *CHM* equation to show how Rayleigh's and Fjortoft's stability theorems are modified by the presence of the background vorticity gradient associated with the differential rotation. We establish linear and nonlinear stability conditions for steady solutions to the model based on the Hamiltonian formalism. In the limit of parallel shear flow solutions, it is shown how the Hamiltonian linear stability results reduce to the modified normal mode Rayleigh's and Fjortoft's stability theorems.

The existence constraints associated with Andrews's Theorem are examined. In particular, we show rigorously that the Arnol'd-like stability theory developed in this section is unable to prove the stability of any steadily-travelling finite-energy and enstrophy solutions in either the plane or a periodic channel domain. This is a serious failure of Hamiltonian stability theory for two dimensional wave and steadily-travelling vortex flows. The problem of developing a mathematical stability theory for steadily-travelling two dimensional flows, based on the Hamiltonian structure, remains an important open research problem.

In Chapter 6 we develop the Hamiltonian structure and associated stability theory for the celebrated *Korteweg-de Vries* or *KdV* equation. The *KdV* equation is a canonical equation in mathematical physics in that it invariably arises whenever one is modelling the nonlinear evolution of small-but-finite and weakly dispersive waves. The *KdV* equation is a nonlinear partial differential equation and was first derived by Korteweg & de Vries (1895) in a study of surface gravity waves.

It was not until the work of Fermi *et al.* (1955) and Zabusky & Kruskal (1965) that evidence began to mount that there was something unusual about solutions to the Cauchy problem associated with the *KdV* equation. In particular, the numerical simulations suggested that under relatively modest assumptions on the initial data, the solution evolved into a finite sequence of amplitude ordered coherent pulses and wave-like radiation. Moreover, these coherent pulses also appeared to have the property that when one pulse interacts with another, the structure of the pulses was preserved (other than a phase shift) after the interaction. This was an extremely remarkable result since the *KdV* equation is *nonlinear* partial differential equation and this property had the appearance of the *principle of superposition* which is a property of *linear* differential equations. This result also suggested that these pulses, which were described as *solitons* by Zabusky & Kruskal (1965), were rather stable structures.

Interest in and the development of a mathematical theory of the *KdV* equation was intensified in the late 1960's and early 1970's. It is not our intention to give a complete overview of all that is known about the *KdV* equation or other soliton equations. The interested reader is referred to, for example, Drazin & Johnson (1989), Newell (1985) or Ablowitz & Segur (1981) for excellent accounts of the theory. For our purposes we pick the story up in 1971 when Gardner (1971) and independently Zakharov & Faddeev (1971) showed that the *KdV* equation was an infinite dimensional Hamiltonian system. Shortly afterward, Benjamin (1972), with additional technical refinements by Bona (1975), established the nonlinear stability in the sense of Liapunov of the *KdV* soliton.

Our goal in Chapter 6 is to present the stability proof for the *KdV* soliton within the context of the *KdV* equation's underlying Hamiltonian structure. The chapter begins with a brief derivation of the *KdV* equation in the context of describing the evolution of weakly nonlinear and dispersive internal gravity waves in a continuously stratified fluid of finite depth under the influence of gravity. We then describe the Hamiltonian formulation and derive the general family of Casimir invariants. The linear momentum invariant is derived as a consequence of applying Noether's Theorem.

It turns out that the Hamiltonian structure of the *KdV* equation is not unique. We briefly describe an alternate Hamiltonian formulation of the *KdV* equation although a detailed verification of this formulation is left as an exercise for the reader.

We then briefly derive the general periodic and soliton solutions of the *KdV* equation. A variational principle is introduced for these solutions in terms a suitably constrained Hamiltonian. The linear stability problem associated with the soliton solution is derived. It is shown that there are no unstable normal mode solutions to the linear stability problem which satisfy the boundary conditions.

The main mathematical problem associated with developing the rigorous (linear and nonlinear) stability theory for the *KdV* soliton is that it is not obvious that the second variation of the functional associated the soliton variational principle (evaluated at the soliton) is definite for all perturbations. By exploiting the spectral properties associated the operator associated with the integrand of the second variation, it is possible to establish appropriate bounds on the second variation and thereby establish stability. We develop this theory with sufficient generality that the reader can see how it applies to other soliton equations.

A final word on the level of presentation. It is assumed that the reader has had some (but not necessarily an exhaustive) exposure to introductory fluid dynamics, hydrodynamic stability theory, finite dimensional Hamiltonian mechanics and variational calculus. As such, this book is written at a level appropriate for senior undergraduate or beginning graduate students in applied mathematics, engineering, physics and specific physical sciences such as fluid and plasma dynamics or dynamical meteorology and oceanography. There is more than enough material for a single semester course. We take a tutorial approach to presenting the subject material.

Chapter 2

The nonlinear pendulum

2.1 Model formulation

The governing equations for the nonlinear pendulum can be written in the form

$$\frac{d^2x}{dt_*^2} = -S\sin(\theta), \tag{2.1}$$

$$\frac{d^2y}{dt_*^2} = -S\cos(\theta) + g, \tag{2.2}$$

where (x, y) are the coordinates of the pendulum as a function of the time t_*, θ is the angle, measured counter clockwise, the pendulum makes with the $y - axis$ as depicted in Fig. 2.1, g is the gravitational acceleration, and S is the tension along the pendulum. Since there is no radial acceleration, it follows that

$$S = g\cos(\theta). \tag{2.3}$$

If the transformation

$$x = L\cos(\theta), \tag{2.4}$$

$$y = L\sin(\theta), \tag{2.5}$$

is introduced into (2.1) and (2.2), where L is the length of the arm of the pendulum, it follows that

$$\frac{d^2\theta}{dt_*^2} + (g/L)\sin(\theta) = 0. \tag{2.6}$$

Introducing the nondimensional time $t = \sqrt{(g/L)}t_*$ into (2.6), leads to the *nondimensional* pendulum equation

$$\theta_{tt} + \sin(\theta) = 0, \tag{2.7}$$

where subscripts will denote the appropriate (partial or ordinary) derivative.

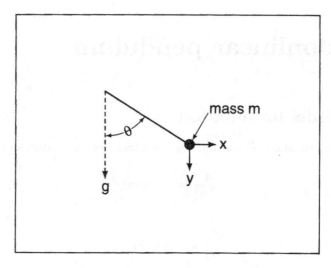

Figure 2.1. Geometry of the pendulum model.

The total energy, denoted by H, is given by

$$H = \frac{\theta_t^2}{2} + [1 - \cos(\theta)],$$

The kinetic and potential energies are given by $\theta_t^2/2$ and $1 - \cos(\theta)$, respectively. The energy is the *Hamiltonian* for this problem. The Hamiltonian is an invariant of the motion since

$$\frac{dH}{dt} = \theta_t\theta_{tt} + \sin(\theta)\theta_t = [\theta_{tt} + \sin(\theta)]\,\theta_t = 0,$$

on account of (2.7).

2.2 Canonical Hamiltonian formulation

The pendulum as a Hamiltonian system

The *canonical* position and momentum are given by, respectively,

$$q = \theta, \tag{2.8}$$

$$p = \theta_t. \tag{2.9}$$

The equation of motion (2.7) can be written in the form

$$q_t = p, \tag{2.10}$$

$$p_t = -\sin(q), \tag{2.11}$$

and the Hamiltonian can be written in the form

$$H(q,p) = \frac{p^2}{2} + [1 - \cos(q)]. \tag{2.12}$$

The governing equations (2.10) and (2.11) can be re-written in the form

$$q_t = \frac{\partial H}{\partial p}, \tag{2.13}$$

$$p_t = -\frac{\partial H}{\partial q}. \tag{2.14}$$

The governing equations (2.13) and (2.14) are the *canonical form of Hamilton's equations* for the pendulum.

General Hamiltonian system

The pendulum corresponds to a two dimensional Hamiltonian system in which there is exactly one generalized position and momentum variable, respectively. In general, there can be n generalized position and momentum variables $\{q_i\}_{i=1}^{i=n}$ and $\{p_i\}_{i=1}^{i=n}$, respectively, giving rise to a $2n$-dimensional Hamiltonian system of the form

$$\frac{dq_i}{dt} = \frac{\partial H(q_j, p_j)}{\partial p_i},$$

$$\frac{dp_i}{dt} = -\frac{\partial H(q_j, p_j)}{\partial q_i},$$

for $i, j = 1, ..., n.$

2.3   Least Action Principle

Least action for a general Hamiltonian system

Hamilton's equations of motion are the *Euler-Lagrange* equations associated with possible extremal solutions to the *action*, denoted $\mathcal{L}$, given by

$$\mathcal{L} = \int_0^T L\left(q_{i_t}, q_i\right) dt,$$

where $L = L\left(q_{i_t}, q_i\right)$ is the *Lagrangian* with $i = 1, ..., n$. The Lagrangian is related to the kinetic and potential energy, KE and PE, respectively, through the relation

$$L = KE - PE.$$

The first variation of the action, denoted as $\delta\mathcal{L}$, is given by

$$\delta\mathcal{L} = \int_0^T \sum_{i=1}^n \left[\frac{\partial L}{\partial q_{i_t}} \delta q_{i_t} + \frac{\partial L}{\partial q_i} \delta q_i \right] dt$$

$$= \int_0^T \sum_{i=1}^n \left\{ \left[-\frac{d}{dt}\left(\frac{\partial L}{\partial q_{i_t}}\right) + \frac{\partial L}{\partial q_i}\right] \delta q_i \right\} dt,$$

where we have integrated by parts once assuming $\delta q_i(0) = \delta q_i(T) = 0$. For an extremal of the action, it follows that $\delta\mathcal{L} = 0$, which implies that

$$\frac{d}{dt}\left(\frac{\partial L}{\partial q_{i_t}}\right) - \frac{\partial L}{\partial q_i} = 0, \ \forall \, i = 1, ..., n,$$

which are the Euler-Lagrange equations associated with the least action principle.

The Hamiltonian is a function of the canonical positions and momenta and *not* q_{i_t} and q_i. The canonical momenta p_i associated with the positions q_i are *defined* by

$$p_i = \frac{\partial L\left(q_{j_t}, q_j\right)}{\partial q_{i_t}},$$

for $i, j = 1, ..., n$. Assuming this relation can be inverted allows one to write a relationship of the form $q_{k_t} = \mathcal{F}(q_m, p_m)$ for $k, m = 1, ..., n$.

The Hamiltonian is related to the Lagrangian through the *Legendre transformation*

$$H\left(q_i, p_i\right) = -L(q_{i_t}, q_i) + \sum_{j=1}^n p_j q_{j_t},$$

where it is understood that $q_{k_t} = \mathcal{F}(q_m, p_m)$ in the right-hand-side of this expression as obtained from inverting the canonical momenta definition.

Equivalently, the Lagrangian is related to the Hamiltonian through the Legendre transformation

$$L(q_{i_t}, q_i) = -H(q_i, p_i) + \sum_{j=1}^{n} p_j q_{j_t},$$

where it is understood that $p_k = \mathcal{G}(q_m, q_{m_t})$ for $k, m = 1, ..., n$ in the right-hand-side of this expression as obtained from inverting the Hamiltonian relation

$$q_{i_t} = \frac{\partial H(q_j, p_j)}{\partial p_i}.$$

Theorem 2.1 *The least action equations*

$$p_i = \frac{\partial L}{\partial q_{i_t}},$$

$$\frac{dp_i}{dt} - \frac{\partial L}{\partial q_i} = 0,$$

for $i = 1, ..., n$ are equivalent to Hamilton's equations of motion.

Proof. Given a Hamiltonian function $H = H(q_i, p_i)$, it follows that

$$dH = \sum_{i=1}^{n} \frac{\partial H}{\partial q_i} dq_i + \frac{\partial H}{\partial p_i} dp_i = \sum_{i=1}^{n} -\frac{dp_i}{dt} dq_i + \frac{dq_i}{dt} dp_i,$$

where we have used Hamilton's equations. However, from the Legendre transformation, it also follows that

$$dH = \sum_{i=1}^{n} \{q_{i_t} dp_i + p_i dq_{i_t} - \frac{\partial L}{\partial q_{i_t}} dq_{i_t} - \frac{\partial L}{\partial q_i} dq_i\}$$

$$= \sum_{i=1}^{n} -\frac{dp_i}{dt} dq_i + \frac{dq_i}{dt} dp_i,$$

if the least action equations are used. Since the two differentials are obviously identical we conclude that the least action and Hamiltonian descriptions of the evolution of the dependent variables are equivalent. ∎

Least action for the pendulum

The action for the pendulum is given by

$$\mathcal{L} = \int\limits_0^T L\left(\theta_t, \theta\right) dt,$$

where the Lagrangian is given by

$$L = \frac{\theta_t^2}{2} + \left[\cos(\theta) - 1\right].$$

The least action equation is therefore simply

$$\frac{d}{dt}\left(\frac{\partial L}{\partial \theta_t}\right) - \frac{\partial L}{\partial \theta} = \theta_{tt} + \sin(\theta) = 0.$$

which just (2.7).

Finally, we can verify that θ_t is indeed the canonical momentum for the pendulum since by direct computation

$$\frac{\partial L}{\partial \theta_t} = \theta_t.$$

The least action principle is a powerful tool in finite dimensional mechanics. Unfortunately, in infinite dimensional mechanics there is often no least action principle available in terms of the Eulerian variables one wants to work with. Thus, rather than having the Hamiltonian structure necessarily follow from a least action principle, it is more convenient to determine the necessary mathematical properties associated with classical Hamiltonian formulations and then *define* a Hamiltonian formulation as a mathematical formulation of the governing equations which satisfies these axioms. In the next section we begin the process of reformulating the canonical Hamiltonian description given in the previous section in such a manner as to facilitate this generalization.

2.4   Symplectic Hamiltonian formulation

Symplectic formulation of the pendulum

The pendulum equations (2.13) and (2.14) can be re-written in the *matrix* form

$$\begin{bmatrix} q \\ p \end{bmatrix}_t = \begin{bmatrix} 0 & 1 \\ -1 & 0 \end{bmatrix} \begin{bmatrix} H_q \\ H_p \end{bmatrix}. \tag{2.15}$$

If the vector notation

$$\mathbf{q} = \begin{bmatrix} q \\ p \end{bmatrix}, \qquad (2.16)$$

$$\mathbf{J} = \begin{bmatrix} 0 & 1 \\ -1 & 0 \end{bmatrix}, \qquad (2.17)$$

$$\frac{\partial H}{\partial \mathbf{q}} = \begin{bmatrix} H_q \\ H_p \end{bmatrix}, \qquad (2.18)$$

is introduced into (2.15), it follows that

$$\mathbf{q}_t = \mathbf{J}\frac{\partial H}{\partial \mathbf{q}}. \qquad (2.19)$$

Equation (2.19) is the *symplectic form of Hamilton's equations* for the pendulum. According to Goldstein (1980), the term symplectic was first introduced in 1939 by H. Weyl from the Greek for "intertwined" to highlight the interdependence of the evolution of the canonical positions and momenta.

General symplectic formulation

The symplectic equations for a general $2n$-dimensional Hamiltonian system consisting of the n position variables $\{q_i\}_{i=1}^{i=n}$ and n momentum variables $\{p_i\}_{i=1}^{i=n}$ can be written in the form (2.19) for

$$\mathbf{q} = \begin{bmatrix} q_1 \\ \vdots \\ q_n \\ p_1 \\ \vdots \\ p_n \end{bmatrix},$$

$$\mathbf{J} = \begin{bmatrix} \mathbf{0} & \mathbf{I} \\ -\mathbf{I} & \mathbf{0} \end{bmatrix},$$

$$\frac{\partial H}{\partial \mathbf{q}} = \begin{bmatrix} H_{q_1} \\ \vdots \\ H_{q_n} \\ H_{p_1} \\ \vdots \\ H_{p_n} \end{bmatrix},$$

where $\mathbf{J}$ is a $2n \times 2n$ matrix with $\mathbf{0}$ and $\mathbf{I}$ the $n \times n$ zero and identity matrices, respectively.

Hamilton's equations of motion for the pendulum represent a *Lagrangian* formulation of the dynamics. That is, they describe the time evolution of the position and momentum of the pendulum following the pendulum. In the pendulum problem, this is convenient since there is only the time evolution of the single "point" corresponding to the tip of the pendulum to be concerned with.

In fluid mechanics the relevant fields are continuous functions of space and it usually more convenient to work with an *Eulerian* formulation of the dynamics which will describe the evolution of fields other than the position and momentum (e.g., the vorticity for the two dimensional Euler equations). Also, because the fields are continuous in fluid mechanics, the evolution of the flow will be described by partial differential equations and not the ordinary differential equations that occur in the pendulum problem.

2.5   Mathematical properties of the J matrix

The phase space associated with the dynamics

The *phase space* associated with a given Hamiltonian system is the symplectic space associated with the dependent variables. For the pendulum, the phase space is the Euclidean space $\mathbb{R}^2$ since there are only the two dependent variables (q, p). For a general finite $2n$-dimensional Hamiltonian system, the phase space will be $\mathbb{R}^{2n}$.

Associated with the phase space is an *inner product,* denoted generically as $\langle \mathbf{a}, \mathbf{b} \rangle$, where $\mathbf{a}$ and $\mathbf{b}$ are arbitrary elements of the phase space. In finite dimensional mechanics, the inner product is the Euclidean inner product. For the pendulum, the inner product is given by

$$\langle \mathbf{a}, \mathbf{b} \rangle = \mathbf{a}^\top \mathbf{b} = a_1 b_1 + a_2 b_2, \tag{2.20}$$

where $\mathbf{a}$ and $\mathbf{b}$ are arbitrary elements of $\mathbb{R}^2$ written as column vectors in the form

$$\mathbf{a} = \left[\begin{array}{c} a_1 \\ a_2 \end{array} \right],$$

$$\mathbf{b} = \left[\begin{array}{c} b_1 \\ b_2 \end{array} \right],$$

and where $(*)^\top$ is the transpose of $(*)$. In finite dimensional mechanics, the phase space *norm* is the Euclidean norm. For the pendulum

$$\|\mathbf{a}\| = \sqrt{\langle \mathbf{a}, \mathbf{a} \rangle},$$

where **a** is an arbitrary element of $\mathbb{R}^2$.

For a general finite $2n$-dimensional Hamiltonian system, the inner product is given by

$$\langle \mathbf{a}, \mathbf{b} \rangle = \mathbf{a}^\top \mathbf{b} = \sum_{i=1}^{2n} a_i b_i, \tag{2.21}$$

where $\mathbf{a}, \mathbf{b} \in \mathbb{R}^{2n}$ and the norm is given by

$$\|\mathbf{a}\| = \left\{ \sum_{i=1}^{2n} a_i b_i \right\}^{\frac{1}{2}}.$$

Skew symmetry of the **J** matrix

A square matrix **M** is called *skew symmetric*, if

$$\langle \mathbf{a}, \mathbf{M}\mathbf{b} \rangle = - \langle \mathbf{M}\mathbf{a}, \mathbf{b} \rangle, \tag{2.22}$$

for all **a** and **b** in the appropriate metric space. The matrix **J** in the symplectic formulation of the pendulum is clearly skew symmetric since

$$\langle \mathbf{q}_1, \mathbf{J}\mathbf{q}_2 \rangle = \mathbf{q}_1^\top \mathbf{J} \mathbf{q}_2 = \mathbf{q}_1^\top \left(-\mathbf{J}^\top \right) \mathbf{q}_2$$

$$= - \left(\mathbf{J}\mathbf{q}_1 \right)^\top \mathbf{q}_2 = - \langle \mathbf{J}\mathbf{q}_1, \mathbf{q}_2 \rangle,$$

for all $\mathbf{q}_1, \mathbf{q}_2 \in \mathbb{R}^2$, where we have used the property that $\mathbf{J} = -\mathbf{J}^\top$.

This latter property of the **J** matrix for the pendulum, i.e.,

$$\mathbf{J} = -\mathbf{J}^\top,$$

is simply the usual definition of skew symmetry appropriate for a matrix whose elements are real numbers. In finite dimensional Hamiltonian dynamics, this is the usual way of expressing the skew symmetry property for the **J** matrix. For infinite dimensional dynamical systems where the elements of **J** are in general partial differential operators, (2.22) is the appropriate formulation of the skew symmetry property.

Noninvertible Hamiltonian dynamics

The matrix **J** for the pendulum is invertible, that is $\mathbf{J}^{-1}$ exists. Specifically,

$$\mathbf{J}^{-1} = -\mathbf{J} = \mathbf{J}^\top.$$

If, for a given Hamiltonian dynamical system $\mathbf{J}^{-1}$ exists, then the Hamiltonian formulation is called *invertible*. If $\mathbf{J}^{-1}$ does not exist, then the Hamiltonian formulation is called *noninvertible*. Finite dimensional Hamiltonian systems can always be formulated with the property that $\mathbf{J}^{-1}$ exists as, for example, the formulation of the pendulum given here. In infinite dimensions, there may not be an invertible formulation in terms of the variables one wants to work with. For example, to best of our current knowledge, the governing equations of fluid dynamics do not have a canonical (or invertible) Hamiltonian formulation in terms of the Eulerian variables (which are the variables of choice). This is very important because the non-invertibility of the $\mathbf{J}$ matrix for the fluid equations is precisely the key to a systematic construction of the pseudo energy needed in the stability results.

2.6   Poisson bracket formulation

It is possible to re-write the dynamical system for the pendulum using a Poisson bracket formulation. Suppose

$$F = F(p, q),$$

$$G = G(p, q),$$

are arbitrary differentiable functions of p and q. The *canonical Poisson bracket* between F and G is defined to be

$$[F, G] = \frac{\partial F}{\partial q}\frac{\partial G}{\partial p} - \frac{\partial F}{\partial p}\frac{\partial G}{\partial q}. \tag{2.23}$$

It will be said that F *Poisson commutes with* G if $[F, G] = 0$.

If the vector notation (2.16), (2.17) and (2.18) is introduced into (2.23), it follows that

$$[F, G] = [F_q \; F_p]\begin{bmatrix} 0 & 1 \\ -1 & 0 \end{bmatrix}\begin{bmatrix} G_q \\ G_p \end{bmatrix} = \left(\frac{\partial F}{\partial \mathbf{q}}\right)^{\top}\mathbf{J}\frac{\partial G}{\partial \mathbf{q}}, \tag{2.24}$$

which can be further simplified, using the inner product (2.21), into the form

$$[F, G] = \left\langle \frac{\partial F}{\partial \mathbf{q}},\ \mathbf{J}\frac{\partial G}{\partial \mathbf{q}} \right\rangle. \tag{2.25}$$

The canonical Poisson bracket for a general $2n$-dimensional Hamiltonian system can be written in the form

$$[F, G] = \sum_{i=1}^{n} \frac{\partial F}{\partial q_i}\frac{\partial G}{\partial p_i} - \frac{\partial F}{\partial p_i}\frac{\partial G}{\partial q_i}, \tag{2.26}$$

where $F = F(p_1, ..., p_n, q_1, ..., q_n)$ and $G = G(p_1, ..., p_n, q_1, ..., q_n)$ are arbitrary functions of the $2n$ independent variables $(p_1, ..., p_n, q_1, ..., q_n)$.

Evolution of a general function

Suppose $F = F(p, q)$ is an arbitrary differentiable function of (p, q) for the pendulum. It follows that

$$\frac{dF}{dt} = \frac{\partial F}{\partial p}\frac{dp}{dt} + \frac{\partial F}{\partial q}\frac{dq}{dt},$$

which can be re-written using (2.13) and (2.14) as

$$\frac{dF}{dt} = \frac{\partial H}{\partial p}\frac{\partial F}{\partial q} - \frac{\partial H}{\partial q}\frac{\partial F}{\partial p},$$

where H is the Hamiltonian (2.12). If the Poisson bracket (2.23) is introduced into this expression, it follows that

$$\frac{dF}{dt} = [F, H]. \tag{2.27}$$

This expression is consistent with (2.13) and (2.14) since it follows from (2.27), respectively, that

$$\frac{dp}{dt} = [p, H] = -\frac{\partial H}{\partial q},$$

$$\frac{dq}{dt} = [q, H] = \frac{\partial H}{\partial p}.$$

It therefore follows from (2.27) that a function $F(p, q)$ is an invariant of the motion if and only if F Poisson commutes with the Hamiltonian H, i.e.,

$$\frac{dF}{dt} = 0 \iff [F, H] = 0.$$

The invariance of the Hamiltonian is obvious since

$$[H, H] = 0.$$

For a general $2n$-dimensional Hamiltonian system with the independent variables $(p_1, ..., p_n, q_1, ..., q_n)$, the evolution of $F(p_1, ..., p_n, q_1, ..., q_n)$ is given by

$$\frac{dF}{dt} = \sum_{i=1}^{n} \frac{\partial F}{\partial q_i}\frac{\partial H}{\partial p_i} - \frac{\partial F}{\partial p_i}\frac{\partial H}{\partial q_i} \equiv [F, H],$$

where the Poisson bracket is understood to be the $2n$-dimensional bracket (2.26).

Algebraic properties of the Poisson bracket

The Poisson bracket satisfies five important algebraic properties. The approach taken here for the presentation of these properties will be based on the symplectic formulation of the Poisson bracket (2.25). This approach is somewhat more compatible with the arguments needed in the infinite dimensional theory to be presented later.

Let $F(\mathbf{q})$, $Q(\mathbf{q})$ and $G(\mathbf{q})$ be arbitrary differentiable functions of $(p_1, ..., p_n, q_1, ..., q_n)$ with $\mathbf{q} = (p_1, ..., p_n, q_1, ..., q_n)^\top$, and let α and β be arbitrary real numbers, it follows that the Poisson bracket satisfies:

P1: Self-Poisson commutation.

$$[F, F] = 0.$$

Proof. It follows from (2.25) and (2.22) that

$$[F, F] = \left\langle \frac{\partial F}{\partial \mathbf{q}}, \mathbf{J}\frac{\partial F}{\partial \mathbf{q}} \right\rangle = -\left\langle \mathbf{J}\frac{\partial F}{\partial \mathbf{q}}, \frac{\partial F}{\partial \mathbf{q}} \right\rangle.$$

However, because the inner product is abelian with respect to its arguments, i.e., $\langle *_1, *_2 \rangle = \langle *_2, *_1 \rangle \; \forall \; *_1$ and $*_2$, it also follows that

$$\left\langle \mathbf{J}\frac{\partial F}{\partial \mathbf{q}}, \frac{\partial F}{\partial \mathbf{q}} \right\rangle = \left\langle \frac{\partial F}{\partial \mathbf{q}}, \mathbf{J}\frac{\partial F}{\partial \mathbf{q}} \right\rangle.$$

Thus we conclude

$$\left\langle \frac{\partial F}{\partial \mathbf{q}}, \mathbf{J}\frac{\partial F}{\partial \mathbf{q}} \right\rangle = -\left\langle \frac{\partial F}{\partial \mathbf{q}}, \mathbf{J}\frac{\partial F}{\partial \mathbf{q}} \right\rangle \implies \left\langle \frac{\partial F}{\partial \mathbf{q}}, \mathbf{J}\frac{\partial F}{\partial \mathbf{q}} \right\rangle = 0. \; \blacksquare$$

P2: Skew-symmetry.

$$[F, G] = -[G, F].$$

Proof. It follows from (2.25), (2.22) and the abelian property of the inner product that

$$[F, G] = \left\langle \frac{\partial F}{\partial \mathbf{q}}, \mathbf{J}\frac{\partial G}{\partial \mathbf{q}} \right\rangle = -\left\langle \mathbf{J}\frac{\partial F}{\partial \mathbf{q}}, \frac{\partial G}{\partial \mathbf{q}} \right\rangle$$

$$= -\left\langle \frac{\partial G}{\partial \mathbf{q}}, \mathbf{J}\frac{\partial F}{\partial \mathbf{q}} \right\rangle = -[G, F]. \; \blacksquare$$

Note that property P2 implies P1.

P3: Distributive property.

$$[\alpha F + \beta G, Q] = \alpha [F, Q] + \beta [G, Q].$$

Proof. It follows from (2.25) that

$$[\alpha F + \beta G, Q] = \left\langle \frac{\partial (\alpha F + \beta G)}{\partial \mathbf{q}}, \ \mathbf{J} \frac{\partial Q}{\partial \mathbf{q}} \right\rangle = \left\langle \alpha \frac{\partial F}{\partial \mathbf{q}} + \beta \frac{\partial G}{\partial \mathbf{q}}, \ \mathbf{J} \frac{\partial Q}{\partial \mathbf{q}} \right\rangle$$

$$= \alpha \left\langle \frac{\partial F}{\partial \mathbf{q}}, \ \mathbf{J} \frac{\partial Q}{\partial \mathbf{q}} \right\rangle + \beta \left\langle \frac{\partial G}{\partial \mathbf{q}}, \ \mathbf{J} \frac{\partial Q}{\partial \mathbf{q}} \right\rangle$$

$$= \alpha [F, Q] + \beta [G, Q]. \ \blacksquare$$

P4: Associative property.

$$[FG, Q] = F [G, Q] + [F, Q] G.$$

Proof. It follows from (2.25) that

$$[FG, Q] = \left\langle \frac{\partial (FG)}{\partial \mathbf{q}}, \ \mathbf{J} \frac{\partial Q}{\partial \mathbf{q}} \right\rangle$$

$$= \left\langle \frac{\partial F}{\partial \mathbf{q}} G + F \frac{\partial G}{\partial \mathbf{q}}, \ \mathbf{J} \frac{\partial Q}{\partial \mathbf{q}} \right\rangle$$

$$= F \left\langle \frac{\partial G}{\partial \mathbf{q}}, \ \mathbf{J} \frac{\partial Q}{\partial \mathbf{q}} \right\rangle + \left\langle \frac{\partial F}{\partial \mathbf{q}}, \ \mathbf{J} \frac{\partial Q}{\partial \mathbf{q}} \right\rangle G$$

$$= F [G, Q] + [F, Q] G,$$

where care has been taken to take account of the possibility that $FG \neq GF$, i.e., F and G need not commute. $\blacksquare$

P5: Jacobi identity.

$$[F, [G, Q]] + [G, [Q, F]] + [Q, [F, G]] = 0.$$

Proof. The proof given here will be valid for a general $2n$-dimensional canonical Hamiltonian formulation. It follows from (2.25) that

$$[F, [G, Q]] + [G, [Q, F]] + [Q, [F, G]]$$

$$= \left[F, \left\langle \frac{\partial G}{\partial \mathbf{q}}, \mathbf{J} \frac{\partial Q}{\partial \mathbf{q}} \right\rangle \right] + \left[G, \left\langle \frac{\partial Q}{\partial \mathbf{q}}, \mathbf{J} \frac{\partial F}{\partial \mathbf{q}} \right\rangle \right] + \left[Q, \left\langle \frac{\partial F}{\partial \mathbf{q}}, \mathbf{J} \frac{\partial G}{\partial \mathbf{q}} \right\rangle \right]$$

$$= \left\langle \frac{\partial F}{\partial \mathbf{q}}, \mathbf{J} \frac{\partial}{\partial \mathbf{q}} \left\langle \frac{\partial G}{\partial \mathbf{q}}, \mathbf{J} \frac{\partial Q}{\partial \mathbf{q}} \right\rangle \right\rangle + \left\langle \frac{\partial G}{\partial \mathbf{q}}, \mathbf{J} \frac{\partial}{\partial \mathbf{q}} \left\langle \frac{\partial Q}{\partial \mathbf{q}}, \mathbf{J} \frac{\partial F}{\partial \mathbf{q}} \right\rangle \right\rangle$$

$$+ \left\langle \frac{\partial Q}{\partial \mathbf{q}}, \mathbf{J} \frac{\partial}{\partial \mathbf{q}} \left\langle \frac{\partial F}{\partial \mathbf{q}}, \mathbf{J} \frac{\partial G}{\partial \mathbf{q}} \right\rangle \right\rangle = - \left\langle \mathbf{J} \frac{\partial F}{\partial \mathbf{q}}, \frac{\partial}{\partial \mathbf{q}} \left\langle \frac{\partial G}{\partial \mathbf{q}}, \mathbf{J} \frac{\partial Q}{\partial \mathbf{q}} \right\rangle \right\rangle$$

$$- \left\langle \mathbf{J} \frac{\partial G}{\partial \mathbf{q}}, \frac{\partial}{\partial \mathbf{q}} \left\langle \frac{\partial Q}{\partial \mathbf{q}}, \mathbf{J} \frac{\partial F}{\partial \mathbf{q}} \right\rangle \right\rangle - \left\langle \mathbf{J} \frac{\partial Q}{\partial \mathbf{q}}, \frac{\partial}{\partial \mathbf{q}} \left\langle \frac{\partial F}{\partial \mathbf{q}}, \mathbf{J} \frac{\partial G}{\partial \mathbf{q}} \right\rangle \right\rangle.$$

To proceed further it is necessary to explicitly compute the three inner product terms in this last expression. The most direct way to compute these terms is write them out using index notation with the summation convention employed. (Henceforth, if double indices occur in a particular term, the summation convention will be assumed unless otherwise stated.) The first inner product can be written in the form

$$\left\langle \mathbf{J} \frac{\partial F}{\partial \mathbf{q}}, \frac{\partial}{\partial \mathbf{q}} \left\langle \frac{\partial G}{\partial \mathbf{q}}, \mathbf{J} \frac{\partial Q}{\partial \mathbf{q}} \right\rangle \right\rangle = J_{ij} \frac{\partial F}{\partial q_j} \frac{\partial}{\partial q_i} \left(\frac{\partial G}{\partial q_k} J_{kl} \frac{\partial Q}{\partial q_l} \right)$$

$$= J_{ij} J_{kl} \frac{\partial F}{\partial q_j} \left(\frac{\partial^2 G}{\partial q_k \partial q_i} \frac{\partial Q}{\partial q_l} + \frac{\partial G}{\partial q_k} \frac{\partial^2 Q}{\partial q_l \partial q_i} \right).$$

The second inner product can be written in the form

$$\left\langle \mathbf{J} \frac{\partial G}{\partial \mathbf{q}}, \frac{\partial}{\partial \mathbf{q}} \left\langle \frac{\partial Q}{\partial \mathbf{q}}, \mathbf{J} \frac{\partial F}{\partial \mathbf{q}} \right\rangle \right\rangle = J_{lk} \frac{\partial G}{\partial q_k} \frac{\partial}{\partial q_l} \left(\frac{\partial Q}{\partial q_i} J_{ij} \frac{\partial F}{\partial q_j} \right)$$

$$= J_{ij} J_{lk} \frac{\partial G}{\partial q_k} \left(\frac{\partial^2 Q}{\partial q_l \partial q_i} \frac{\partial F}{\partial q_j} + \frac{\partial Q}{\partial q_i} \frac{\partial^2 F}{\partial q_l \partial q_j} \right).$$

The third inner product can be written in the form

$$\left\langle \mathbf{J} \frac{\partial Q}{\partial \mathbf{q}}, \frac{\partial}{\partial \mathbf{q}} \left\langle \frac{\partial F}{\partial \mathbf{q}}, \mathbf{J} \frac{\partial G}{\partial \mathbf{q}} \right\rangle \right\rangle = J_{mn} \frac{\partial Q}{\partial q_n} \frac{\partial}{\partial q_m} \left(\frac{\partial F}{\partial q_u} J_{uv} \frac{\partial G}{\partial q_v} \right)$$

$$= J_{mn} J_{uv} \frac{\partial Q}{\partial q_n} \frac{\partial G}{\partial q_v} \frac{\partial^2 F}{\partial q_m \partial q_u} + J_{mn} J_{uv} \frac{\partial Q}{\partial q_n} \frac{\partial F}{\partial q_u} \frac{\partial^2 G}{\partial q_m \partial q_v}.$$

There are two terms in this expression. If the transformations

$$(m, n, u, v) \rightarrow (j, i, l, k),$$

$$(m, n, u, v) \rightarrow (k, l, j, i),$$

are introduced into these last two terms, respectively, it follows that

$$\left\langle \mathbf{J}\frac{\partial Q}{\partial \mathbf{q}}, \frac{\partial}{\partial \mathbf{q}}\left\langle \frac{\partial F}{\partial \mathbf{q}}, \mathbf{J}\frac{\partial G}{\partial \mathbf{q}}\right\rangle\right\rangle = J_{ji}J_{lk}\frac{\partial Q}{\partial q_i}\frac{\partial G}{\partial q_k}\frac{\partial^2 F}{\partial q_l \partial q_j}$$

$$+ J_{kl}J_{ji}\frac{\partial Q}{\partial q_l}\frac{\partial F}{\partial q_j}\frac{\partial^2 G}{\partial q_k \partial q_i}.$$

If these representations are substituted back in for the inner products, it follows that

$$[F, [G, Q]] + [G, [Q, F]] + [Q, [F, G]]$$

$$= (J_{ij} + J_{ji})J_{kl}\frac{\partial F}{\partial q_j}\frac{\partial Q}{\partial q_l}\frac{\partial^2 G}{\partial q_k \partial q_i}$$

$$+ (J_{kl} + J_{lk})J_{ij}\frac{\partial G}{\partial q_k}\frac{\partial F}{\partial q_j}\frac{\partial^2 Q}{\partial q_l \partial q_i}$$

$$+ (J_{lk} + J_{kl})J_{ji}\frac{\partial Q}{\partial q_i}\frac{\partial G}{\partial q_k}\frac{\partial^2 F}{\partial q_l \partial q_j} = 0,$$

since $\mathbf{J}^\top = -\mathbf{J}$. ∎

One important point to make about this proof of the Jacobi identity is that we have explicitly exploited the underlying canonical structure, that is we have

$$\frac{\partial J_{ij}}{\partial q_n} = 0,$$

for all (i, j, n). This fact significantly simplifies the verification of the Jacobi identity. In the noncanonical Hamiltonian formulation to be developed for the Euler equations, the elements of $\mathbf{J}$ will *explicitly* depend on the dependent variables $\mathbf{q}$ and the verification of the Jacobi identity is somewhat more delicate.

Transformation rule for J and the Poisson bracket

In some problems it may be of advantage to work with a new set of *dependent* variables $\phi_k = \phi_k(q_l)$. In this section we will determine the transformation rules for the Poisson bracket and the **J** matrix. It will be assumed that $\phi_k = \phi_k(q_l)$ is a diffeomorphism (i.e., ϕ_k is one-to-one and ϕ_k and its inverse are continuously differentiable).

We will denote the Hamiltonian written with respect to the ϕ_k variables as $\widetilde{H} = \widetilde{H}(\phi_k)$, i.e.,

$$H\left(q_l\right) = \widetilde{H}\left(\phi_k\left(q_l\right)\right).$$

The transformation rule for **J** can be derived by examining the time evolution of ϕ_k, which is given by

$$\frac{d\phi_k}{dt} = \frac{\partial\phi_k}{\partial q_i}\frac{dq_i}{dt} = \frac{\partial\phi_k}{\partial q_i}J_{ij}\frac{\partial H}{\partial q_j}$$

$$= \frac{\partial\phi_k}{\partial q_i}J_{ij}\frac{\partial\widetilde{H}}{\partial\phi_m}\frac{\partial\phi_m}{\partial q_j}$$

$$= \widetilde{J}_{km}\frac{\partial\widetilde{H}}{\partial\phi_m},$$

where

$$\widetilde{J}_{km} \equiv \frac{\partial\phi_k}{\partial q_i}J_{ij}\frac{\partial\phi_m}{\partial q_j}.$$

Thus we have shown that **J** transforms as a rank-2 contravariant tensor.

The Poisson bracket transforms similarly. Let $F = F(q_i)$ and $G = G(q_i)$ be arbitrary functions of the **q** variables with $\widetilde{F} = \widetilde{F}(\phi_k)$ and $\widetilde{G} = \widetilde{G}(\phi_k)$ the corresponding functions with respect to the ϕ variables, i.e.,

$$F(q_i) = \widetilde{F}(\phi_k(q_i)),$$

$$G(q_i) = \widetilde{G}(\phi_k(q_i)).$$

If we denote the Poisson bracket between F and G written with respect to the **q** variables as $[F,G]_{\mathbf{q}}$ and the Poisson bracket between $\widetilde{F}$ and $\widetilde{G}$ written with respect to the ϕ variables as $\left[\widetilde{F}, \widetilde{G}\right]_{\phi}$, it follows that

$$[F,G]_{\mathbf{q}} \equiv \frac{\partial F}{\partial q_i}J_{ij}\frac{\partial G}{\partial q_j}$$

$$= \frac{\partial \widetilde{F}}{\partial \phi_k} \frac{\partial \phi_k}{\partial q_i} J_{ij} \frac{\partial \widetilde{G}}{\partial \phi_m} \frac{\partial \phi_m}{\partial q_j}$$

$$= \frac{\partial \widetilde{F}}{\partial \phi_k} \widetilde{J}_{ij} \frac{\partial \widetilde{G}}{\partial \phi_m} \equiv \left[\widetilde{F}, \widetilde{G} \right]_\phi.$$

Casimir functions

A *Casimir* is a function of the dependent variables that Poisson commutes with every other function of the dependent variables. That is, $C = C(\mathbf{q})$ is a Casimir if

$$[F, C] = 0, \ \forall \ F(\mathbf{q}). \tag{2.28}$$

The first thing to observe is that a Casimir is obviously an invariant of the motion since

$$\frac{dC}{dt} = [H, C] = 0,$$

on account of (2.28).

The second thing to observe is that it follows from (2.25) that (2.28) is equivalent to

$$\left\langle \frac{\partial F}{\partial \mathbf{q}}, \mathbf{J} \frac{\partial C}{\partial \mathbf{q}} \right\rangle = 0, \ \forall \ F(\mathbf{q}).$$

However, since $F(\mathbf{q})$ and hence $\frac{\partial F}{\partial \mathbf{q}}$ is arbitrary, it follows from the properties of inner products that

$$\mathbf{J} \frac{\partial C}{\partial \mathbf{q}} = \mathbf{0}. \tag{2.29}$$

Thus the set of Casimirs can be characterized as those functions $C = C(\mathbf{q})$ such that

$$\frac{\partial C}{\partial \mathbf{q}} \in \ker(\mathbf{J}).$$

If $\mathbf{J}^{-1}$ exists, then it follows from (2.29) that

$$\frac{\partial C}{\partial \mathbf{q}} = \mathbf{0} \Longrightarrow C = C_0,$$

where C_0 is a constant. In this situation the Casimirs are said to be *trivial*. On the other hand, if $\mathbf{J}^{-1}$ does *not* exist, then the kernel of $\mathbf{J}$ contains at least one nonzero element and there will exist at least one nontrivial Casimir. Since $\mathbf{J}^{-1}$ exists for the

Hamiltonian structure given for the pendulum there are only trivial Casimirs in that formulation.

Invertible and noninvertible Hamiltonian formulations can therefore be distinguished from each other with the statement

> invertible Hamiltonian formulation ⟺ trivial Casimirs.

It will be shown that a systematic construction of Arnol'd's pseudo energy centres on identifying the Casimirs associated with a noninvertible Hamiltonian formulation of the relevant fluid dynamics equations.

If a given Hamiltonian formulation is noninvertible, then the equations of motion are invariant under the addition of a Casimir to the Hamiltonian function. This can be easily seen as follows. Suppose $C = C(\mathbf{q})$ is a Casimir associated with the symplectic equation

$$\mathbf{q}_t = \mathbf{J}\frac{\partial H}{\partial \mathbf{q}},$$

then

$$\mathbf{J}\frac{\partial (H+C)}{\partial \mathbf{q}} = \mathbf{J}\frac{\partial H}{\partial \mathbf{q}} + \mathbf{J}\frac{\partial C}{\partial \mathbf{q}} = \mathbf{q}_t + 0 = \mathbf{q}_t.$$

2.7 Steady solutions of a canonical Hamiltonian system

Variational principle for a general steady solution

A steady or equilibrium solution to a general Hamiltonian system is a solution of the form

$$\mathbf{q} = \mathbf{q}_o, \tag{2.30}$$

which time independent, i.e.,

$$\frac{d\mathbf{q}_o}{dt} = 0. \tag{2.31}$$

Substituting (2.30) into (2.19) yields

$$\mathbf{J}\frac{\partial H}{\partial \mathbf{q}}(\mathbf{q}_o) = 0 \iff \frac{\partial H}{\partial \mathbf{q}}(\mathbf{q}_o) = 0, \tag{2.32}$$

assuming $\mathbf{J}^{-1}$ exists. Steady solutions for an invertible Hamiltonian system therefore satisfy the first order necessary conditions for a local extremum of the energy. It is possible to give the following variational principle for a general steady solution.

Theorem 2.2 *Every steady solution $\mathbf{q}_o$ of an invertible Hamiltonian system satisfies the variational principle*

$$dH(\mathbf{q}_o) = 0,$$

where $dH(\mathbf{q}_o)$ is the total differential of the Hamiltonian $H(\mathbf{q})$ evaluated at $\mathbf{q} = \mathbf{q}_o$.

Proof. Direct calculation of dH gives

$$dH(\mathbf{q}) = \frac{\partial H}{\partial q_i} dq_i = \left\langle \frac{\partial H}{\partial \mathbf{q}}, d\mathbf{q} \right\rangle.$$

Thus, if $\mathbf{q} = \mathbf{q}_o$ satisfies

$$\frac{\partial H}{\partial \mathbf{q}}(\mathbf{q}_o) = 0,$$

it follows that

$$dH(\mathbf{q}_o) = 0. \quad \blacksquare$$

Corollary 2.3 *Steady solutions of a noninvertible Hamiltonian system do not necessarily satisfy the variational principle given in Theorem 2.2 above.*

Proof. Let $\mathbf{q}_o$ be a steady solution of the Hamiltonian system

$$\frac{d\mathbf{q}}{dt} = \mathbf{J} \frac{\partial H}{\partial \mathbf{q}},$$

i.e.,

$$\mathbf{J} \frac{\partial H}{\partial \mathbf{q}}(\mathbf{q}_o) = 0.$$

However, since the Hamiltonian formulation is noninvertible, it follows that $\mathbf{J}^{-1}$ does not exist. Therefore it cannot be concluded that

$$\frac{\partial H}{\partial \mathbf{q}}(\mathbf{q}_o) = 0. \quad \blacksquare$$

Steady solutions of the pendulum

Since the Hamiltonian formulation of the pendulum is invertible, it follows that all steady solutions will satisfy the variational principle given in Theorem 2.2. Substituting the pendulum Hamiltonian (2.12) into (2.32) yields the two equations

$$p_o = 0, \tag{2.33}$$

$$\sin(q_o) = 0. \tag{2.34}$$

These are, of course, equivalent to the equations obtained for steady solutions from the original governing equations (2.10) and (2.11). From (2.34) we conclude

$$q_o = n\pi, \; n = 0, \pm 1, \pm 2, \dots . \tag{2.35}$$

The steady solutions given by the even and odd multiples of π correspond, physically, to the pendulum, as depicted in Fig. 2.1, being stationary and oriented downward and upward, respectively.

2.8   Linear stability of a steady solution

Linear stability equations

The general linear stability equations associated with a steady solution q_o can be obtained in the following way. Suppose the *total* solution $q(t)$ can be written as a sum of the steady solution q_o and a perturbation term $\delta q(t)$ in the form

$$q = q_o + \delta q, \tag{2.36}$$

where it is implicitly assumed that δq is infinitesimally small in the sense that

$$0 < \|\delta q\| \ll 1,$$

to the degree that the evolution of δq can be determined (at least for some period of time) by a linearized approximation to the full nonlinear governing equations.

If (2.36) is substituted into

$$\frac{dq}{dt} = J\frac{\partial H}{\partial q},$$

and the Hamiltonian Taylor expanded about q_o, the result can be written in the form

$$\frac{d}{dt}\delta q_i = J_{ij}\left[\frac{\partial H}{\partial q_j}(q_o) + \frac{\partial^2 H}{\partial q_j \partial q_k}(q_o)\delta q_k + h.o.t.\right],$$

where *h.o.t.* stands for *higher order terms*. Neglecting the higher order terms and recognizing that

$$\frac{\partial H}{\partial q_j}(q_o) = 0,$$

for $j = 1, \dots, 2n$, respectively, leads to the *linear stability equations* in the form

$$\frac{d}{dt}\delta q_i = J_{ij}\frac{\partial^2 H}{\partial q_j \partial q_k}(q_o)\delta q_k, \tag{2.37}$$

for $i = 1, \dots, 2n$, respectively. The issue of the linear stability of q_o revolves around whether or not, roughly speaking, the norm of δq increases in magnitude as time increases.

Definition 2.4 *The equilibrium solution* $\mathbf{q}_o$ *is said to be linearly stable in the sense of Liapunov if for every* $\varepsilon > 0$ *there exists* $\delta > 0$ *such that* $\|\boldsymbol{\delta q}\| < \varepsilon$ *for all* $t \geq 0$ *whenever* $\|\boldsymbol{\delta q}\|_{t=0} < \delta$, *where* $\boldsymbol{\delta q}$ *solves the linear stability equations (2.37). If* $\mathbf{q}_o$ *is not linearly stable it is called linearly unstable.*

If a steady solution is linearly stable in the sense of Liapunov, then both algebraic *and* normal mode instabilities (i.e., when the perturbation amplitude grows algebraically or exponentially, respectively, with respect to time) are ruled out.

Linear stability equations for the pendulum

Substituting the pendulum Hamiltonian (2.12) into (2.37) yields the linear stability equations

$$\delta q_t = \delta p, \qquad (2.38)$$

$$\delta p_t = -\cos(q_o)\delta q = (-1)^{n+1}\delta q, \qquad (2.39)$$

where (2.35) has been used. These two equations can be combined together to give

$$\delta q_{tt} + (-1)^n \delta q = 0. \qquad (2.40)$$

Clearly, if n is odd, then there are exponentially growing solutions as t increases. If n is even, there are only sinusoidal solutions which remain bounded for all time. We expect, therefore, that the steady solution corresponding to the pendulum being stationary and oriented downward (that is, n being even) is *linearly stable* and the other steady solution corresponding to the pendulum being stationary but oriented upward (that is, n being odd) is *unstable*.

The details of the mathematical argument are straightforward. If n is even, the general solution for δq and δp is given by

$$\delta q = \alpha \sin(t) + \beta \cos(t),$$

$$\delta p = \alpha \cos(t) - \beta \sin(t),$$

where α and β are arbitrary constants. It follows that

$$\|\boldsymbol{\delta q}\| = \sqrt{(\delta q)^2 + (\delta p)^2} = \sqrt{\alpha^2 + \beta^2} = \|\boldsymbol{\delta q}\|_{t=0}.$$

It therefore follows that $\|\boldsymbol{\delta q}\|_{t=0} < \varepsilon \implies \|\boldsymbol{\delta q}\| < \varepsilon, \ \forall \ t \geq 0$ and the linear stability in the sense of Liapunov of the steady solutions is established if n is even.

If n is odd, the general solution for δq and δp is given by

$$\delta q = \alpha \sinh(t) + \beta \cosh(t),$$

$$\delta p = \alpha \cosh(t) - \beta \sinh(t),$$

where α and β are arbitrary constants. It follows that

$$\|\boldsymbol{\delta q}\| = \sqrt{(\delta q)^2 + (\delta p)^2} = \sqrt{(\alpha^2 + \beta^2)\cosh(2t)}$$

$$= \|\boldsymbol{\delta q}\|_{t=0} \sqrt{\cosh(2t)} \geq \frac{1}{\sqrt{2}} \exp(t) \|\boldsymbol{\delta q}\|_{t=0}.$$

Consequently for every choice of $\|\boldsymbol{\delta q}\|_{t=0} \neq 0$

$$\lim_{t\to\infty} \|\boldsymbol{\delta q}\| = \infty,$$

and therefore $\mathbf{q}_o$ is clearly unstable if n is odd.

General linear stability theory

The general linear stability equations for a constant coefficient $2n$-dimensional Hamiltonian system can always be solved exactly. However, it is *not* generally possible to solve the *infinite dimensional* linear stability equations that arise in fluid mechanics. An alternate approach is therefore needed for determining the general stability characteristics of steady solutions to Hamiltonian systems. Consequently, we now develop a stability theory for canonical Hamiltonian systems which does not require explicitly solving the linear stability equations.

It has been shown that steady solutions of an invertible Hamiltonian system satisfy the first order necessary conditions for an extremum of the energy. Consider now the second differential of H, i.e.,

$$d^2 H(\mathbf{q}) = \frac{\partial^2 H}{\partial q_i \partial q_j}(\mathbf{q}) dq_i dq_j + \frac{\partial H}{\partial q_i}(\mathbf{q}) d^2 q_i.$$

It follows that

$$d^2 H(\mathbf{q}_o) = \frac{\partial^2 H}{\partial q_i \partial q_j}(\mathbf{q}_o) dq_i dq_j. \tag{2.41}$$

The second differential of the Hamiltonian evaluated at the steady solution is obviously a quadratic form with a symmetric coefficient matrix (assuming, as we do, that H is at least a twice continuously differentiable function of its arguments).

Theorem 2.5 *The quadratic form*

$$d^2 H(\mathbf{q}_o) = \frac{\partial^2 H}{\partial q_i \partial q_j}(\mathbf{q}_o)\delta q_i \delta q_j,$$

in an invariant of the linear stability equations

$$\frac{d}{dt}\delta q_i = J_{ij}\frac{\partial^2 H}{\partial q_j \partial q_k}(\mathbf{q}_o)\delta q_k.$$

Proof. We show that

$$\frac{d}{dt}\{\frac{\partial^2 H}{\partial q_i \partial q_j}(\mathbf{q}_o)\delta q_i \delta q_j\} = 0.$$

We may write

$$\frac{d}{dt}\{\frac{\partial^2 H}{\partial q_i \partial q_j}(\mathbf{q}_o)\delta q_i \delta q_j\} = (\delta q_i)_t \frac{\partial^2 H}{\partial q_i \partial q_j}(\mathbf{q}_o)\delta q_j$$

$$+ \delta q_i \frac{\partial^2 H}{\partial q_i \partial q_j}(\mathbf{q}_o)(\delta q_j)_t$$

$$= \delta q_k \frac{\partial^2 H}{\partial q_k \partial q_l}(\mathbf{q}_o) J_{il} \frac{\partial^2 H}{\partial q_i \partial q_j}(\mathbf{q}_o)\delta q_j$$

$$+ \delta q_i \frac{\partial^2 H}{\partial q_i \partial q_j}(\mathbf{q}_o) J_{jl} \frac{\partial^2 H}{\partial q_l \partial q_k}(\mathbf{q}_o)\delta q_k$$

$$= 2\delta q_k \frac{\partial^2 H}{\partial q_k \partial q_l}(\mathbf{q}_o) J_{il} \frac{\partial^2 H}{\partial q_i \partial q_j}(\mathbf{q}_o)\delta q_j, \qquad (2.42)$$

since the terms are identical.

If Ω is defined to be the vector with components given by

$$\Omega_i \equiv \frac{\partial^2 H}{\partial q_i \partial q_j}(\mathbf{q}_o)\delta q_j,$$

for $i = 1, ..., 2n$, then (2.42) can be re-written in the form

$$\frac{d}{dt}\{\frac{\partial^2 H}{\partial q_i \partial q_j}(\mathbf{q}_o)\delta q_i \delta q_j\} = 2\langle \Omega, \mathbf{J}\Omega \rangle.$$

However, since $\mathbf{J}$ is skew symmetric and the inner product abelian, it follows that

$$\langle \Omega, \mathbf{J}\Omega \rangle = -\langle \mathbf{J}\Omega, \Omega \rangle = -\langle \Omega, \mathbf{J}\Omega \rangle \implies \langle \Omega, \mathbf{J}\Omega \rangle = 0.$$

Which implies

$$\frac{d}{dt}\{\frac{\partial^2 H}{\partial q_i \partial q_j}(\mathbf{q}_o)\delta q_i \delta q_j\} = 0. \quad \blacksquare$$

The invariance of the second differential of the Hamiltonian evaluated at the steady solution with respect to the linear stability equations is an important property because it will be proven that if $d^2 H(\mathbf{q}_o)$ is positive or negative definite for all (nontrivial) perturbations $\boldsymbol{\delta q}$, then $\mathbf{q}_o$ is linearly stable in the sense of Liapunov.

The definiteness of (2.41) has a simple geometrical interpretation:

$$\frac{\partial^2 H}{\partial q_i \partial q_j}(\mathbf{q}_o)\delta q_i \delta q_j > 0, \ \forall \ \boldsymbol{\delta q} \neq 0 \Longleftrightarrow H(\mathbf{q}_o) \ is \ a \ local \ \min,$$

$$\frac{\partial^2 H}{\partial q_i \partial q_j}(\mathbf{q}_o)\delta q_i \delta q_j < 0, \ \forall \ \boldsymbol{\delta q} \neq 0 \Longleftrightarrow H(\mathbf{q}_o) \ is \ a \ local \ \max.$$

The general procedure for determining whether or not (2.41) is definite involves examining the signs of the *principal minors* of the *Hessian* matrix associated with H, denoted as $[\mathcal{H}_{ij}]$, defined by

$$\mathcal{H}_{ij} \equiv \frac{\partial^2 H}{\partial q_i \partial q_j}(\mathbf{q}_o).$$

The principal minors, denoted by $M_1, M_2, M_3, ..., M_{2n}$, are given by

$$M_1 = \mathcal{H}_{11},$$

$$M_2 = \det \begin{bmatrix} \mathcal{H}_{11} & \mathcal{H}_{12} \\ \mathcal{H}_{21} & \mathcal{H}_{22} \end{bmatrix},$$

$$M_3 = \det \begin{bmatrix} \mathcal{H}_{11} & \mathcal{H}_{12} & \mathcal{H}_{13} \\ \mathcal{H}_{21} & \mathcal{H}_{22} & \mathcal{H}_{23} \\ \mathcal{H}_{31} & \mathcal{H}_{32} & \mathcal{H}_{33} \end{bmatrix},$$

and so on until

$$M_{2n} = \det \begin{bmatrix} \mathcal{H}_{11} & \cdots & \mathcal{H}_{1,2n} \\ \vdots & \ddots & \vdots \\ \mathcal{H}_{2n,1} & \cdots & \mathcal{H}_{2n,2n} \end{bmatrix}.$$

The determinantal test for the definiteness of $d^2 H(\mathbf{q}_o)$ can therefore be stated in the form

$$M_i > 0, \ \forall \ i = 1, ..., 2n \Longleftrightarrow d^2 H(\mathbf{q}_o) > 0, \qquad (2.43)$$

$$(-1)^i M_i > 0, \ \forall \ i = 1, ..., 2n \Longleftrightarrow d^2 H(\mathbf{q}_o) < 0. \qquad (2.44)$$

Theorem 2.6 *If*

$$d^2 H(\mathbf{q}_o) = \frac{\partial^2 H}{\partial q_i \partial q_j}(\mathbf{q}_o) \delta q_i \delta q_j,$$

is definite for all $\delta\mathbf{q}$, then the steady solution $\mathbf{q}_o$ is linearly stable in the sense of Liapunov.

Proof. There are two cases that need to be examined. First, suppose

$$d^2 H(\mathbf{q}_o) = \frac{\partial^2 H}{\partial q_i \partial q_j}(\mathbf{q}_o) \delta q_i \delta q_j > 0, \ \forall \ \delta\mathbf{q} \neq \mathbf{0}. \tag{2.45}$$

Since the Hessian $[\mathcal{H}_{ij}]$ is a real symmetric matrix, it follows from (2.45) that the eigenvalues are all strictly positive. Let $\lambda_{\min}$ and $\lambda_{\max}$, with the ordering

$$0 < \lambda_{\min} \leq \lambda_{\max} < \infty,$$

be the minimum and maximum eigenvalues of $[\mathcal{H}_{ij}]$, respectively. It follows from the theory of quadratic forms that

$$0 < \lambda_{\min}||\delta\mathbf{q}||^2 \leq \frac{\partial^2 H}{\partial q_i \partial q_j}(\mathbf{q}_o) \delta q_i \delta q_j \leq \lambda_{\max}||\delta\mathbf{q}||^2,$$

for all nontrivial perturbations. Thus, exploiting the invariance of $d^2 H(\mathbf{q}_o)$, we find that

$$0 < \lambda_{\min}||\delta\mathbf{q}||^2 \leq \frac{\partial^2 H}{\partial q_i \partial q_j}(\mathbf{q}_o) \delta q_i \delta q_j$$

$$= \frac{\partial^2 H}{\partial q_i \partial q_j}(\mathbf{q}_o)[\delta q_i \delta q_j]_{t=0} \leq \lambda_{\max}||\delta\mathbf{q}||_{t=0}^2,$$

which implies

$$0 < ||\delta\mathbf{q}|| \leq \left[\frac{\lambda_{\max}}{\lambda_{\min}}\right]^{\frac{1}{2}} ||\delta\mathbf{q}||_{t=0}.$$

Consequently, for every $\varepsilon > 0$,

$$0 < ||\delta\mathbf{q}||_{t=0} < \varepsilon \left[\frac{\lambda_{\min}}{\lambda_{\max}}\right]^{\frac{1}{2}} \implies 0 < ||\delta\mathbf{q}|| < \varepsilon, \ \forall \ t \geq 0,$$

and thus the linear stability of $\mathbf{q}_o$ is established if (2.45) holds.

In the case where

$$d^2 H(\mathbf{q}_o) = \frac{\partial^2 H}{\partial q_i \partial q_j}(\mathbf{q}_o) \delta q_i \delta q_j < 0, \ \forall \ \delta \mathbf{q} \neq \mathbf{0}, \tag{2.46}$$

the theory of real symmetric matrices guarantees that the eigenvalues of the Hessian matrix $[\mathcal{H}_{ij}]$ are all strictly negative. Again, let $\lambda_{\min}$ and $\lambda_{\max}$, with the ordering

$$-\infty < \lambda_{\min} \leq \lambda_{\max} < 0,$$

be the minimum and maximum eigenvalues of $[\mathcal{H}_{ij}]$, respectively. As before, from the theory of quadratic forms we have

$$\lambda_{\min} ||\delta \mathbf{q}||^2 \leq \frac{\partial^2 H}{\partial q_i \partial q_j}(\mathbf{q}_o) \delta q_i \delta q_j \leq \lambda_{\max} ||\delta \mathbf{q}||^2 < 0,$$

for all nontrivial perturbations. Thus, exploiting the invariance of $d^2 H(\mathbf{q}_o)$, we find that

$$\lambda_{\min} ||\delta \mathbf{q}||^2_{t=0} \leq \frac{\partial^2 H}{\partial q_i \partial q_j}(\mathbf{q}_o) [\delta q_i \delta q_j]_{t=0}$$

$$= \frac{\partial^2 H}{\partial q_i \partial q_j}(\mathbf{q}_o) \delta q_i \delta q_j \leq \lambda_{\max} ||\delta \mathbf{q}||^2 < 0,$$

which implies

$$0 < ||\delta \mathbf{q}|| \leq \left[\frac{\lambda_{\min}}{\lambda_{\max}}\right]^{\frac{1}{2}} ||\delta \mathbf{q}||_{t=0}.$$

Consequently, for every $\varepsilon > 0$,

$$0 < ||\delta \mathbf{q}||_{t=0} < \varepsilon \left[\frac{\lambda_{\max}}{\lambda_{\min}}\right]^{\frac{1}{2}} \implies 0 < ||\delta \mathbf{q}|| < \varepsilon, \ \forall \ t \geq 0,$$

and thus the linear stability of $\mathbf{q}_o$ is established if (2.46) holds. Consequently, we conclude that if $d^2 H(\mathbf{q}_o)$ is definite for all nontrivial $\delta \mathbf{q}$, then $\mathbf{q}_o$ is linearly stable in the sense of Liapunov. ∎

The general theory applied to the pendulum

Let us re-examine the stability of the steady solutions to the pendulum in the context of Theorem 2.6. The 2×2 Hessian matrix will be given by

$$\mathcal{H}_{ij} \equiv \begin{bmatrix} \mathcal{H}_{qq} & \mathcal{H}_{qp} \\ \mathcal{H}_{pq} & \mathcal{H}_{pp} \end{bmatrix} = \begin{bmatrix} \cos(\mathbf{q}_o) & 0 \\ 0 & 1 \end{bmatrix} = \begin{bmatrix} (-1)^n & 0 \\ 0 & 1 \end{bmatrix}.$$

Thus, if n is even, then

$$d^2 H(\mathbf{q}_o) = \mathcal{H}_{ij} \delta q_i \delta q_j = (\delta q)^2 + (\delta p)^2 > 0,$$

provided $(\delta q, \delta p) \neq \mathbf{0}$, implying the linear stability of the even n solutions.

If n is odd, then

$$d^2 H(\mathbf{q}_o) = \mathcal{H}_{ij} \delta q_i \delta q_j = -(\delta q)^2 + (\delta p)^2,$$

which is indefinite and so the linear stability of the odd n solutions cannot be established.

2.9   Nonlinear stability of a steady solution

The argument just presented is a proof, based on the second differential being definite, of the *linear* stability of an equilibrium solution to a canonical Hamiltonian system. While linear stability is an important result, the desirable result is to the establish *nonlinear* stability of an equilibrium solution to a general Hamiltonian system of the form

$$\mathbf{q}_t = \mathbf{J} \frac{\partial H}{\partial \mathbf{q}}. \tag{2.47}$$

Definition 2.7 *The steady solution $\mathbf{q}_o$ to (2.47) is said to be nonlinearly stable in the sense of Liapunov if for every $\varepsilon > 0$ there exists $\delta > 0$ such that $\|\mathbf{q} - \mathbf{q}_o\| < \varepsilon$ for all $t \geq 0$ whenever $\|\mathbf{q} - \mathbf{q}_o\|_{t=0} < \delta$, where $\mathbf{q}(t)$ solves the full nonlinear system of equations (2.47). If $\mathbf{q}_o$ is not nonlinearly stable, then it is unstable.*

Nonlinear stability differs from linear stability in that the *perturbation*

$$\delta \mathbf{q} \equiv \mathbf{q} - \mathbf{q}_o,$$

is no longer assumed to be infinitesimally small in magnitude. As it turns out, in finite dimensions, the definiteness of $d^2 H(\mathbf{q}_o)$ is sufficient to establish nonlinear stability. This is a consequence of the compactness of finite dimensional vector spaces. The analogue of the second differential in infinite dimensions is the second variation and, as we shall see, the definiteness of the second variation is not sufficient to prove nonlinear stability. The problem is that, whereas in finite dimensions the definiteness of the second differential implies that a local extremum occurs at the steady solution, in infinite dimensions the definiteness of the second variation does not imply that a local extremum occurs at the steady solution. In terms of the mathematical details of proving nonlinear stability in finite versus infinite dimensions, this is the most important consequence of the differing topologies.

General nonlinear stability theory

Theorem 2.8 *The steady solution* q_o *of the Hamiltonian system (2.47) is nonlinearly stable in the sense of Liapunov if* $d^2 H(q_o)$ *is definite.*

Proof. Let $\mathcal{N}(\delta q)$ be the function defined by

$$\mathcal{N}(\delta q) = H(q_o + \delta q) - H(q_o) - dH(q_o), \qquad (2.48)$$

where

$$q \equiv q_o + \delta q,$$

solves the full nonlinear system (2.47). Clearly, $\mathcal{N}(\delta q)$ is an invariant of the full nonlinear equations since $H(q)$ is exactly the Hamiltonian, $H(q_o)$ is completely time independent, and $dH(q_o) = 0$.

If $\mathcal{N}(\delta q)$ is Taylor expanded about 0, it follows that

$$\mathcal{N}(\delta q) = \frac{1}{2} \frac{\partial^2 H}{\partial q_i \partial q_j}(q_o) \delta q_i \delta q_j + \frac{1}{6} \frac{\partial^3 H}{\partial q_i \partial q_j \partial q_k}(q_o) \delta q_i \delta q_j \delta q_k + \dots$$

$$= \frac{1}{2} d^2 H(q_o) + h.o.t. \,,$$

where it is assumed that H and hence $\mathcal{N}$ are smooth functions of their arguments. It follows from the *Mean Value Theorem* that there exists $\xi \in (0, 1)$ so that

$$\mathcal{N}(\delta q) = \frac{1}{2} \frac{\partial^2 H}{\partial q_i \partial q_j}(q_o + \xi \delta q) \delta q_i \delta q_j.$$

Let us first suppose that $d^2 H(q_o) > 0$. But this implies that $H(q_o)$ is a local minimum, i.e., there exists $\mu > 0$ such that

$$H(q_o + \delta q) > H(q_o), \text{ whenever } 0 < \|\delta q\| < \mu.$$

Thus, from the Mean Value Theorem, there exists $\xi \in (0, 1)$ so that

$$\mathcal{N}(\delta q) = \frac{1}{2} \frac{\partial^2 H}{\partial q_i \partial q_j}(q_o + \xi \delta q) \delta q_i \delta q_j > 0,$$

for each perturbation satisfying $0 < \|\delta q\| < \mu$. Consequently, the theory of real symmetric matrices guarantees that the eigenvalues of

$$\frac{\partial^2 H}{\partial q_i \partial q_j}(q_o + \xi \delta q),$$

are all strictly positive.

Let $\lambda_{min}(\delta \mathbf{q})$ and $\lambda_{max}(\delta \mathbf{q})$, with the ordering

$$0 < \lambda_{min} \le \lambda_{max} < \infty,$$

be the minimum and maximum eigenvalues of this matrix, respectively, for a given $\delta \mathbf{q}$. It follows from the assumed smoothness of $\mathcal{N}(\delta \mathbf{q})$ that the eigenvalues will be continuous functions of their arguments. Therefore, there exist *positive constants* α_1 and β_1 such that

$$0 < \alpha_1 < \lambda_{min}(\delta \mathbf{q}) \le \lambda_{max}(\delta \mathbf{q}) < \beta_1 < \infty,$$

for all $0 < ||\delta \mathbf{q}|| < \mu$. It therefore follows, from the theory of quadratic forms, that the convexity estimate

$$0 < \alpha_1 ||\delta \mathbf{q}||^2 \le \frac{\partial^2 H}{\partial q_i \partial q_j}(\mathbf{q}_o + \xi \delta \mathbf{q}) \delta q_i \delta q_j \le \beta_1 ||\delta \mathbf{q}||^2 < \infty,$$

holds for all nontrivial perturbations satisfying $0 < ||\delta \mathbf{q}|| < \mu$.

Thus, exploiting the invariance of $\mathcal{N}(\delta \mathbf{q})$, we obtain

$$0 < \alpha_1 ||\delta \mathbf{q}||^2 \le 2\mathcal{N}(\delta \mathbf{q}) = 2\mathcal{N}(\delta \mathbf{q})|_{t=0} \le \beta_1 ||\delta \mathbf{q}||^2_{t=0} < \infty,$$

and conclude

$$0 < ||\delta \mathbf{q}|| \le \left[\frac{\beta_1}{\alpha_1}\right]^{\frac{1}{2}} ||\delta \mathbf{q}||_{t=0},$$

provided $0 < ||\delta \mathbf{q}|| < \mu$.

Therefore, given $\varepsilon > 0$, there exists δ given by

$$0 < \delta = \min(\varepsilon, \mu) \left[\frac{\alpha_1}{\beta_1}\right]^{\frac{1}{2}} \le \mu,$$

such that

$$||\delta \mathbf{q}||_{t=0} < \delta \le \mu \Longrightarrow ||\delta \mathbf{q}|| < \min(\varepsilon, \mu) \le \varepsilon, \ \forall \ t \ge 0,$$

which establishes the nonlinear stability of $\mathbf{q}_o$ if $d^2 H(\mathbf{q}_o) > 0$.

The case supposing $d^2 H(\mathbf{q}_o) < 0$, i.e., $H(\mathbf{q}_o)$ is a local maximum, proceeds in a completely analogous manner. There exists $\mu > 0$ such that

$$H(\mathbf{q}_o + \delta \mathbf{q}) < H(\mathbf{q}_o), \ whenever \ 0 < ||\delta \mathbf{q}|| < \mu.$$

Thus, from the Mean Value Theorem, there exists $\xi \in (0, 1)$ for which

$$\mathcal{N}(\delta \mathbf{q}) = \frac{1}{2}\frac{\partial^2 H}{\partial q_i \partial q_j}(\mathbf{q}_o + \xi \delta \mathbf{q}) \delta q_i \delta q_j < 0,$$

for each perturbation satisfying $0 < ||\delta\mathbf{q}|| < \mu$. Consequently, the theory of real symmetric matrices guarantees that the eigenvalues of

$$\frac{\partial^2 H}{\partial q_i \partial q_j}(\mathbf{q}_o + \xi\delta\mathbf{q}),$$

are all strictly negative.

Let $\lambda_{\min}(\delta\mathbf{q})$ and $\lambda_{\max}(\delta\mathbf{q})$, with the ordering

$$-\infty < \lambda_{\min} \leq \lambda_{\max} < 0,$$

be the minimum and maximum eigenvalues of this matrix, respectively. As before, the continuity of the eigenvalues implies the existence of *negative constants* α_2 and β_2 such that

$$-\infty < \alpha_2 < \lambda_{\min}(\delta\mathbf{q}) \leq \lambda_{\max}(\delta\mathbf{q}) < \beta_2 < 0,$$

for all $0 < ||\delta\mathbf{q}|| < \mu$. It therefore follows, from the theory of quadratic forms, that the convexity estimate

$$-\infty < \alpha_2||\delta\mathbf{q}||^2 \leq \frac{\partial^2 H}{\partial q_i \partial q_j}(\mathbf{q}_o + \xi\delta\mathbf{q})\delta q_i \delta q_j \leq \beta_2||\delta\mathbf{q}||^2 < 0,$$

holds for all nontrivial perturbations satisfying $0 < ||\delta\mathbf{q}|| < \mu$.

Thus, exploiting the invariance of $\mathcal{N}(\delta\mathbf{q})$, we obtain

$$-\infty < \alpha_2||\delta\mathbf{q}||^2_{t=0} \leq 2\mathcal{N}(\delta\mathbf{q})|_{t=0} = 2\mathcal{N}(\delta\mathbf{q}) \leq \beta_2||\delta\mathbf{q}||^2 < 0,$$

and conclude

$$0 < ||\delta\mathbf{q}|| \leq \left[\frac{\alpha_2}{\beta_2}\right]^{\frac{1}{2}} ||\delta\mathbf{q}||_{t=0},$$

provided $0 < ||\delta\mathbf{q}|| < \mu$.

Therefore, given $\varepsilon > 0$, there exists δ given by

$$0 < \delta = \min(\varepsilon, \mu)\left[\frac{\beta_2}{\alpha_2}\right]^{\frac{1}{2}} \leq \mu,$$

such that

$$||\delta\mathbf{q}||_{t=0} < \delta \leq \mu \Longrightarrow ||\delta\mathbf{q}|| < \min(\varepsilon, \mu) \leq \varepsilon, \ \forall\, t \geq 0,$$

which establishes the nonlinear stability of $\mathbf{q}_o$ if $d^2 H(\mathbf{q}_o) < 0$. Consequently, we conclude that if $d^2 H(\mathbf{q}_o)$ is definite, then $\mathbf{q}_o$ is nonlinearly stable in the sense of Liapunov. ∎

The general theory applied to the pendulum

Let us re-examine the stability of the steady solutions to the pendulum in the context of Theorem 2.8. We will show, as expected, that the steady solutions (2.33) and (2.34) are nonlinearly stable if n is even and nonlinear stability cannot be established if n is odd.

If the pendulum Hamiltonian (2.12) is substituted into (2.48), it follows that

$$\mathcal{N}(\delta q, \delta p) = \frac{(p_o + \delta p)^2}{2} + 1 - \cos(q_o + \delta q)$$

$$- \left[\frac{(p_o)^2}{2} + 1 - \cos(q_o) \right] - [p_o \delta p + \sin(q_o)\, \delta q]$$

$$= \frac{(\delta p)^2}{2} + (-1)^n [1 - \cos(\delta q)], \tag{2.49}$$

where $(q_o, p_o) = (n\pi, 0)$ has been substituted in. Because of the periodicity of $\cos(*)$ we may, without loss of generality, restrict attention to $-\pi < \delta q \leq \pi$ in (2.49).

We need to determine nonzero convexity constants α and β so that

$$\alpha\, (\delta q)^2 \leq 1 - \cos(\delta q) \leq \beta\, (\delta q)^2. \tag{2.50}$$

The Mean Value Theorem asserts the existence of q_* between 0 and δq so that

$$1 - \cos(\delta q) = \frac{\cos(q_*)(\delta q)^2}{2}. \tag{2.51}$$

Since $\cos(q_*) \leq 1$ we have

$$1 - \cos(\delta q) \leq \frac{(\delta q)^2}{2},$$

and thus we may choose

$$\beta = \frac{1}{2}. \tag{2.52}$$

The lower bound takes a little more work. Let us introduce the function $h(x)$ defined by

$$h(x) = \begin{cases} x^{-2}[1 - \cos(x)] & \text{if } x \neq 0, \\ 0.5 & \text{if } x = 0, \end{cases}$$

(see Fig. 2.2). Because $h(x)$ is an even function it will suffice to only consider the interval $x \in [0, \pi]$. Clearly, $h(x)$ is continuously differentiable on the open interval

$x \in (0, \pi)$ and continuous on the right at $x = 0$ and on the left at $x = \pi$ and is monotonically decreasing on $(0, \pi)$; see Fig. 2.2. It therefore follows that α is given by

$$\alpha = h(\pi) = \frac{2}{\pi^2} \simeq 0.20264 > 0.$$

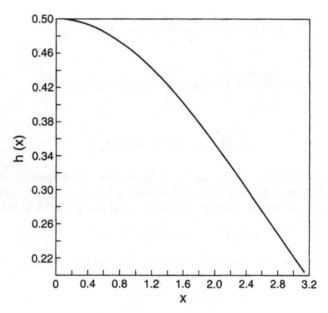

Figure 2.2. Graph of the function $h(x)$.

We may rigorously establish the monotonic behavior of $h(x)$ for $x \in (0, \pi)$ via an argument based on the Mean Value Theorem. Direct computation of $\dfrac{dh}{dx}$ yields

$$h'(x) = \frac{x \sin(x) + 2[\cos(x) - 1]}{x^3}.$$

To show that $h(x)$ is monotonically decreasing all we need show is that $h'(x) \leq 0$ for $x \in (0, \pi)$. Since the denominator of this expression is positive in the open interval all we need show is that the numerator is negative. Let $f(x)$ be given by

$$f(x) = x \sin(x) + 2[\cos(x) - 1].$$

Clearly,

$$f(0) = 0,$$

$$f'(0) = 0,$$

$$f''(x) = -x \sin(x).$$

It follows from the Mean Value Theorem that there exists $\lambda \in (0, x)$ such that

$$f(x) - f(0) - f'(0)x = \frac{f''(\lambda)}{2}x^2.$$

Substitution of $f(x)$ into this expression implies

$$x \sin(x) + 2[\cos(x) - 1] = -\frac{\lambda \sin(\lambda)}{2}x^2.$$

However, since

$$\frac{\lambda \sin(\lambda)}{2}x^2 \geq 0,$$

for all $0 \leq \lambda \leq x \leq \pi$ we conclude that $h'(x) \leq 0$ and thus $h(x) \geq h(\pi)$ for all $x \in (0, \pi)$, i.e.,

$$1 - \cos(x) \geq \frac{2x^2}{\pi^2},$$

for all $x \in (0, \pi)$. Hence we may choose

$$\alpha = \frac{2}{\pi^2}. \tag{2.53}$$

Substituting (2.52) and (2.53) into (2.50) gives the convexity estimate

$$\frac{2(\delta q)^2}{\pi^2} \leq 1 - \cos(\delta q) \leq \frac{(\delta q)^2}{2}, \tag{2.54}$$

for all $\delta q \in (-\pi, \pi]$.

Let us return to (2.49) in the case where n is even. It follows from the invariance of $\mathcal{N}(\delta q, \delta p)$ and the convexity estimate (2.54) that

$$\frac{2}{\pi^2}||\delta q||^2 = \frac{2}{\pi^2}\left[(\delta p)^2 + (\delta q)^2\right]$$

$$\leq \frac{(\delta p)^2}{2} + 1 - \cos(\delta q) = \mathcal{N}(\delta q, \delta p)$$

$$= \mathcal{N}(\delta q, \delta p)|_{t=0} = \left[\frac{(\delta p)^2}{2} + 1 - \cos(\delta q)\right]_{t=0}$$

$$\leq \frac{1}{2}\left[(\delta p)^2 + (\delta q)^2\right]_{t=0} = \frac{1}{2}\|\delta \mathbf{q}\|_{t=0}^2,$$

which implies that

$$\|\delta \mathbf{q}\|(t) \leq \frac{\pi}{2}\|\delta \mathbf{q}\|_{t=0}.$$

Consequently, given $\varepsilon > 0$, it follows that

$$\|\delta \mathbf{q}\|_{t=0} < \frac{2}{\pi}\min[\varepsilon, \pi] \implies \|\delta \mathbf{q}\|(t) < \min[\varepsilon, \pi] < \varepsilon, \ \forall \, t \geq 0,$$

and thus the *nonlinear* stability of the even n steady solutions has been established.

The nonlinear stability of the odd n steady solutions cannot be established since the convexity estimate (2.54) implies that

$$\frac{(\delta p)^2 - (\delta q)^2}{2} \leq \mathcal{N}(\delta q, \delta p) \leq \frac{(\delta p)^2}{2} - \frac{2\,(\delta q)^2}{\pi^2},$$

which is indefinite.

Chapter 3

The two-dimensional Euler equations

3.1 Vorticity equation formulation

The two dimensional Euler equations for a homogeneous fluid can be written in the form

$$u_t + uu_x + vu_y = -\frac{1}{\rho}p_x, \tag{3.1}$$

$$v_t + uv_x + vv_y = -\frac{1}{\rho}p_y, \tag{3.2}$$

$$u_x + v_y = 0, \tag{3.3}$$

where the velocity is given by $\mathbf{u} = (u, v)$, and the pressure and density are given by p and ρ, respectively. The spatial variables are (x, y) and t is time. Subscripts indicate the appropriate partial derivative unless otherwise stated. Equations (3.1) and (3.2) are the momentum equations in the x and y directions, respectively, and (3.3) expresses conservation of mass.

It follows from (3.3) that there exists a scalar *stream function*, denoted $\psi(x, y, t)$, which is related to the velocity components through

$$u = -\psi_y, \tag{3.4}$$

$$v = \psi_x. \tag{3.5}$$

These relations may be expressed in the vector notation

$$\mathbf{u} = (-\psi_y, \psi_x) = \mathbf{e}_3 \times \nabla \psi \equiv \begin{vmatrix} \mathbf{e}_1 & \mathbf{e}_2 & \mathbf{e}_3 \\ 0 & 0 & 1 \\ \psi_x & \psi_y & 0 \end{vmatrix}, \tag{3.6}$$

where the third component in $\mathbf{e}_3 \times \nabla \psi$ is ignored since it is always zero.

The vorticity equation, obtained by forming

$$\partial_x(3.2) - \partial_y(3.1),$$

i.e., the *curl* of the momentum equations, is given by

$$(v_x - u_y)_t + u(v_x - u_y)_x + v(v_x - u_y)_y = 0, \tag{3.7}$$

where mass conservation has been used. This expression states that the vorticity, given by

$$v_x - u_y = \mathbf{e}_3 \cdot [\nabla \times \mathbf{u}],$$

is conserved following the motion. If the stream function relations (3.4) and (3.5) are substituted into (3.7), it follows that the vorticity equation can be written in the form

$$\Delta \psi_t + \partial(\psi, \Delta \psi) = 0, \tag{3.8}$$

where $\Delta = \partial_{xx} + \partial_{yy}$, and

$$\partial(A, B) \equiv A_x B_y - A_y B_x. \tag{3.9}$$

The quantity $\partial(A, B)$ which can be interpreted as the determinate of the Jacobian matrix of A and B with respect to x and y, i.e.,

$$\partial(\dot{A}, B) = \begin{vmatrix} A_x & A_y \\ B_x & B_y \end{vmatrix},$$

will be referred to as simply the *Jacobian between A and B*. It will be necessary, from time to time, to use one of the following alternate ways of expressing the Jacobian:

$$\partial(A, B) = \nabla \cdot [B\mathbf{e}_3 \times \nabla(A)], \tag{3.10}$$

$$= -\nabla \cdot [A\mathbf{e}_3 \times \nabla(B)], \tag{3.11}$$

$$= \mathbf{e}_3 \cdot [\nabla(A) \times \nabla(B)], \tag{3.12}$$

where $\nabla = (\partial_x, \partial_y)$. It is left as an exercise for the reader to verify these formulae.

Thus, the solution to the Euler equations can be obtained by solving the vorticity equation (3.8) for the stream function (subject to, of course, appropriate initial and boundary conditions). The velocity field will be given by (3.6). The pressure is determined from the *Poisson* equation

$$\Delta p = -2\rho[u_y v_x + (u_x)^2],$$

obtained from the divergence of the momentum equations, i.e.,

$$\partial_x(3.1) + \partial_y(3.2),$$

subject to, of course, the appropriate boundary condition (given later in this section). Substitution of (3.6) into this relation implies that the pressure is related to the stream function through

$$\Delta p = 2\rho[\psi_{xx}\psi_{yy} - (\psi_{xy})^2]. \tag{3.13}$$

Thus, we see that Δp is proportional to the Gaussian curvature of the stream function, denoted here as $\mathcal{W}$, i.e.,

$$\mathcal{W} = [\psi_{xx}\psi_{yy} - (\psi_{xy})^2].$$

Weiss (1991) has shown that $\mathcal{W}$ is the trace of the square of the stress tensor. When $\mathcal{W} > 0$ the local flow field is elliptical and when $\mathcal{W} < 0$ the local flow field is hyperbolic. When the local flow field is hyperbolic, one expects vorticity advection to dominate and destroy the local spatial structure through shear. When the local flow is elliptical, one expects vortical motion to dominate.

Since $\mathcal{W} > 0$ (local elliptical structure) implies $\Delta p > 0$, it follows that one expects strong vortical motion to be correlated with a local pressure deficit. Indeed, flow visualization of coherent eddy or vortex structures is sometimes done by presenting isosurfaces of local pressure anomalies (e.g., Robinson, 1991; see also Holmes *et al.*, 1996).

Algebraic properties of the Jacobian

Let $A(x, y, t)$, $B(x, y, t)$ and $C(x, y, t)$ be arbitrary smooth functions of their arguments and let α and β be arbitrary real numbers, then the Jacobian satisfies the properties:

J1: Self commutation.

$$\partial(A, A) = 0.$$

Proof. It follows from (3.9) that

$$\partial(A, A) = A_x A_y - A_y A_x = 0. \ \blacksquare$$

J2: Skew symmetry.

$$\partial(A, B) = -\partial(B, A).$$

Proof. It follows from (3.10) and (3.11) that

$$\partial(A, B) = \nabla \cdot [B\mathbf{e_3} \times \nabla(A)]$$

$$= -\nabla \cdot [A\mathbf{e_3} \times \nabla(B)] = -J(B, A). \ \blacksquare$$

J3: Distributive property.

$$\partial(\alpha A + \beta B, C) = \alpha\partial(A, C) + \beta\partial(B, C).$$

Proof. It follows from (3.10) that

$$\partial(\alpha A + \beta B, C) = \nabla \cdot \{[\mathbf{e}_3 \times \nabla (\alpha A + \beta B) C]\}$$

$$= \alpha\nabla \cdot [C\mathbf{e}_3 \times \nabla (A)] + \beta\nabla \cdot [C\mathbf{e}_3 \times \nabla (B)]$$

$$= \alpha\partial(A, C) + \beta\partial(B, C). \blacksquare$$

J4: Associative property.

$$\partial(AB, C) = B\partial(A, C) + A\partial(B, C).$$

Proof. It follows from (3.12) that

$$\partial(AB, C) = \mathbf{e}_3 \cdot [\nabla (AB) \times \nabla (C)]$$

$$= \mathbf{e}_3 \cdot [B\nabla (A) \times \nabla (C)] + \mathbf{e}_3 \cdot [A\nabla (B) \times \nabla(C)]$$

$$= B\mathbf{e}_3 \cdot [\nabla(A) \times \nabla(C)] + A\mathbf{e}_3 \cdot [\nabla(B) \times \nabla(C)]$$

$$= B\partial(A, C) + A\partial(B, C). \blacksquare$$

J5: "Jacobi identity".

$$\partial(A, \partial(B, C)) + \partial(B, \partial(C, A)) + \partial(C, \partial(A, B)) = 0.$$

Proof. We proceed directly. It follows from (3.9) that

$$\partial(A, \partial(B, C)) + \partial(B, \partial(C, A)) + \partial(C, \partial(A, B))$$

$$= A_x \left(\partial(B, C)\right)_y - A_y \left(\partial(B, C)\right)_x$$

$$+ B_x \left(\partial(C, A)\right)_y - B_y \left(\partial(C, A)\right)_x$$

$$+ C_x \left(\partial(A, B) \right)_y - C_y \left(\partial(A, B) \right)_x$$

$$= A_x \left(B_x C_y - B_y C_x \right)_y - A_y \left(B_x C_y - B_y C_x \right)_x$$

$$+ B_x \left(C_x A_y - C_y A_x \right)_y - B_y \left(C_x A_y - C_y A_x \right)_x$$

$$+ C_x \left(A_x B_y - A_y B_x \right)_y - C_y \left(A_x B_y - A_y B_x \right)_x$$

$$= \left[A_x \left(B_x C_y - B_y C_x \right) \right]_y - \left[A_y \left(B_x C_y - B_y C_x \right) \right]_x$$

$$+ \left[B_x \left(C_x A_y - C_y A_x \right) \right]_y - \left[B_y \left(C_x A_y - C_y A_x \right)_x \right]_x$$

$$+ \left[C_x \left(A_x B_y - A_y B_x \right) \right]_y - \left[C_y \left(A_x B_y - A_y B_x \right) \right]_x$$

$$= \left[A_x B_x C_y - A_x B_x C_y \right]_y + \left[A_x B_y C_x - A_x B_y C_x \right]_y$$

$$+ \left[A_y B_x C_x - A_y B_x C_x \right]_y + \left[A_y B_y C_x - A_y B_y C_x \right]_x$$

$$+ \left[A_y B_x C_y - A_y B_x C_y \right]_x + \left[A_x B_y C_y - A_x B_y C_y \right]_x = 0. \ \blacksquare$$

These properties will be needed later in our presentation of the Hamiltonian structure for the vorticity equation (3.8).

Boundary conditions

We assume that the flow occurs on a fixed *spatial* domain defined to be the open set $\Omega \subseteq \mathbb{R}^2$. This domain may be bounded or unbounded (in one or all directions) and it may or may not be simply connected (see Fig. 3.1). The fixed boundary of Ω, if it exists, will be denoted $\partial\Omega$. If the domain is multiply connected, it is assumed that $\partial\Omega$ is the union of a finite number of smooth simply connected boundary curves $\{\partial\Omega_1, ..., \partial\Omega_n\}$ so that

$$\partial\Omega = \bigcup_{i=1}^{n} \partial\Omega_i,$$

(see Fig. 3.2).

Since the Euler equations (3.1) and (3.2) are inviscid, the appropriate boundary condition is the *no-normal mass flux* condition given by

$$[\mathbf{n} \cdot \mathbf{u}]_{\partial\Omega} = 0, \tag{3.14}$$

where $\mathbf{n}$ is the unit outward *normal* vector on $\partial\Omega$.

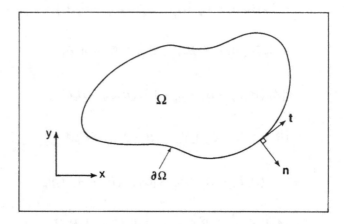

Figure 3.1. An example of a simply connected bounded domain.

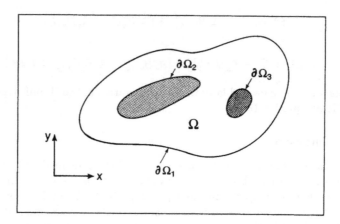

Figure 3.2. An example of a multiply connected bounded domain.

This boundary condition can be expressed in terms of the stream function as follows. Suppose that the normal is written in the component form $\mathbf{n} = (n_1, n_2)$ (where $n_1(x, y)$ and $n_2(x, y)$ satisfy $n_1^2 + n_2^2 = 1$), it follows that on $\partial\Omega$

$$\mathbf{n} \cdot \mathbf{u} = n_1 u + n_2 v = -n_1 \psi_y + n_2 \psi_x$$

$$= (n_2, -n_1) \cdot \nabla \psi = -\mathbf{t} \cdot \nabla \psi \equiv -\frac{\partial \psi}{\partial s},$$

where $\mathbf{t} = (-n_2, n_1)$ is the (right-handed) unit *tangent* vector on $\partial \Omega$ and s is the arclength. Thus, the no-normal mass flux condition implies

$$\left[\frac{\partial \psi}{\partial s} \right]_{\partial \Omega} = 0,$$

which means that

$$\psi = \lambda_i, \ i = 1, ..., n, \tag{3.15}$$

where λ_i is a constant, for *each* smooth simply connected boundary curve $\partial \Omega_i$ of the boundary $\partial \Omega$. If Ω is bounded *and* simply connected, then we may, without loss of generality, choose

$$\psi = 0 \ on \ \partial \Omega, \tag{3.16}$$

since the stream function is only defined up to an additive constant. If the domain Ω is unbounded the appropriate boundary conditions will be specified as they are needed.

The boundary condition for the pressure problem (3.13) can be formulated as follows. If

$$[n_1(3.1) + n_2(3.2)]_{\partial \Omega},$$

is formed, we obtain

$$[\mathbf{n} \cdot \nabla p]_{\partial \Omega} \equiv \left[\frac{\partial p}{\partial n} \right]_{\partial \Omega} = -\rho [\mathbf{n} \cdot (\mathbf{u} \cdot \nabla) \mathbf{u}]_{\partial \Omega},$$

which can be re-written in terms of the stream function as

$$\left[\frac{\partial p}{\partial n} \right]_{\partial \Omega} = \rho \left[\mathbf{n} \cdot \left(\psi_x \psi_{yy} - \psi_y \psi_{xy}, \psi_y \psi_{xx} - \psi_x \psi_{xy} \right) \right]_{\partial \Omega}.$$

The Poisson equation (3.13) together with this *Neumann* boundary condition uniquely determines (up to an additive constant) the pressure field $p(x, y, t)$ in terms of the stream function as determined from the vorticity equation (3.8) and boundary condition (3.16).

Existence and uniqueness of solutions

There is a long history (see, for example, Wolibner, 1933) in connection with the study of the existence, uniqueness and regularity of solutions to the Cauchy problem associated with the two dimensional incompressible Euler equations. It is, however, beyond the scope of this book to properly address these issues and their resolution. Here, we shall simply comment that the global existence of a unique classical solution to the Cauchy problem associated with the two dimensional incompressible Euler equations in a bounded domain, which can be written in the form

$$\mathbf{u}_t + (\mathbf{u} \cdot \nabla) \mathbf{u} = -\frac{1}{\rho} \nabla p,$$

$$\nabla \cdot \mathbf{u} = 0,$$

where $\mathbf{u} = (u, v)$ for $(x, y, t) \in \Omega \times [0, \infty)$, with the boundary conditions

$$\mathbf{n} \cdot \mathbf{u}|_{\partial \Omega} = 0, \quad \forall \quad t \geq 0,$$

and initial condition

$$\mathbf{u}(x, y, 0) = \mathbf{u}_0(x, y),$$

$$\nabla \cdot \mathbf{u}_0 = 0,$$

can be proven via functional analytic fixed point theory, provided the initial data is sufficiently smooth (see Kato, 1967 and earlier work on weak solutions to the above Cauchy problem by Judovic, 1966).

By the unique solution, we mean, of course, that the pressure can only be known up to an arbitrary additive function of time. If the domain is unbounded, a similar global existence theorem is also provable provided the additional hypothesis that the area-integrated energy is finite holds. The reader who is interested in pursuing these issues further is referred to Marchioro & Pulvirenti (1994) and the references contained therein.

3.2   Hamiltonian structure for partial differential equations

In many situations of interest, a system of partial differential equations does not have a least action principle associated with it in terms of the desired dependent variables. For example, while the Euler equations (3.1), (3.2) and (3.3) have a variational principle in terms of Lagrangian variables (e.g., Virasoro, 1981; Salmon, 1988), they do *not* have a variational principle in terms of the Eulerian variables u, v and p (Seliger & Whitham, 1968; Atherton & Homsy, 1975) unless a Clebsch transformation is introduced (see the exercises). It is, therefore, not possible to derive a Hamiltonian structure for the Euler equations as a consequence of an underlying least action principle and accompanying Legendre transformation in terms of the variables we want to work with.

It is, however, possible to *define* a Hamiltonian formulation of a general system of partial differential equations as a mathematical formulation which satisfies the algebraic properties associated with a classical Hamiltonian system as described in Sections 2.5 and 2.6. To the best of our knowledge, the first attempt to provide a noncanonical Hamiltonian formulation of the equations of fluid dynamics (and ideal magnetohydrodynamics) was given by Morrison & Greene (1980). The definition given here closely parallels the approach taken by, for example, Morrison (1982), Olver (1982) or Benjamin (1984).

Definition of a Hamiltonian system of partial differential equations

A system of n partial differential equations written in the form

$$\mathcal{F}(\mathbf{q}, \frac{\partial}{\partial x_i}, \frac{\partial}{\partial t}) = \mathbf{0}, \tag{3.17}$$

where t is time and $\mathbf{q}(\mathbf{x}, t) = (q_1(\mathbf{x}, t), ..., q_n(\mathbf{x}, t))^{\top}$ is a column vector of n dependent variables with the m independent spatial variables $\mathbf{x} = (x_1, ..., x_m)$ defined on the open spatial domain $\Omega \subseteq \mathbb{R}^m$ with the boundary (if it exists) $\partial\Omega$, is said to be *Hamiltonian* if there exists a *conserved functional* $H(\mathbf{q})$, called the Hamiltonian, and a matrix $\mathbf{J}$ of (possibly pseudo) differential operators such that (3.17) can be written in the form

$$\mathbf{q}_t = \mathbf{J}\frac{\delta H}{\delta \mathbf{q}}, \tag{3.18}$$

where $\dfrac{\delta H}{\delta \mathbf{q}}$ is the vector variational or Euler-Lagrange (see Gelfand & Fomin, 1963) derivative of H with respect to $\mathbf{q}$. In addition, the *bracket* defined by

$$[F, G] \equiv \left\langle \frac{\delta F}{\delta \mathbf{q}}, \mathbf{J}\frac{\delta G}{\delta \mathbf{q}} \right\rangle, \tag{3.19}$$

where F and G are arbitrary smooth functionals of $\mathbf{q}$ (where the inner product is typically the $\mathbf{L}^2$ inner product) *must* satisfy the properties:

H1: Self commutation.

$$[F, F] = 0, \tag{3.20}$$

H2: Skew symmetry.

$$[F, G] = -[G, F], \tag{3.21}$$

H3: Distributive property.

$$[\alpha F + \beta G, Q] = \alpha[F, Q] + \beta[G, Q], \tag{3.22}$$

H4: Associative property.

$$[FG, Q] = F[G, Q] + [F, Q]G, \tag{3.23}$$

H5: Jacobi identity.

$$[F, [G, Q]] + [G, [Q, F]] + [Q, [F, G]] = 0, \tag{3.24}$$

where Q is an arbitrary smooth functional of $\mathbf{q}$ and α and β are arbitrary real numbers. If the bracket (3.19) satisfies the five properties *H1, H2, H3, H4* and *H5*, the we call the bracket a *Poisson bracket*. We will say that F *Poisson commutes with* G if $[F, G] = 0$.

It is usually the case that the first four of these properties, i.e., (3.20), (3.21), (3.22) and (3.23), are relatively easy to satisfy in comparison with the Jacobi identity. The Jacobi identity puts a strong constraint on the structure of the J matrix and usually the most difficult to verify.

One question which immediately arises is: How does one determine the J matrix (assuming it exists) for a given system of partial differential equations? The method that seems to be most often used is simply inspired guess work based on previous results for similar problems. As far as the author is aware, there is no general theory presently available for determining whether or not a given system of partial differential equations admits a Hamiltonian formulation.

However, if a given system of partial differential equations is associated with a physical problem for which there exists a *canonical* Hamiltonian description written in terms of the canonical positions and momenta, it may be possible to introduce a sequence of transformations into the canonical Poisson bracket and systematically derive the Poisson bracket as a function of the desired noncanonical dependent variables. If it is possible to determine the Poisson bracket in this way, then J can be determined from (3.19). In Section (3.3) we will show how the Poisson bracket and hence J associated with the vorticity equation (3.8) can be derived by a systematic reduction of the canonical Poisson bracket for the two dimensional Euler equations.

Functionals and variational derivatives

Before continuing on with our treatment of the Euler equations we include a very brief aside on the definition of a functional and the practical calculation of a variational derivative of a smooth functional. For a complete discussion of these topics the interested reader is referred, for example, to Courant & Hilbert (1962) or Gelfand & Fomin (1963). The reader who is already familiar with these concepts may skip this subsection.

A functional, denoted by F, is an integral expression of the form

$$F(\mathbf{q}) = \int_{\Omega} \mathcal{F}(\mathbf{q})d\mathbf{x}, \tag{3.25}$$

where $\mathcal{F}(\mathbf{q})$ is the *density* associated with $F(\mathbf{q})$ and where the volume differential has been written in the form $d\mathbf{x} \equiv dx_1 dx_2 ... dx_m$. The density will be assumed to be a smooth function of its argument(s) unless otherwise stated. Although we have written the argument of the density in (3.25) as $\mathbf{q}$ alone, it is understood that F may depend explicitly on the derivatives of $\mathbf{q}$, the spatial variables or time.

The *vector variational* or *Euler derivative* of a functional is a natural extension of the concept of a partial derivative. The variational derivative of $F(\mathbf{q})$ with respect to $\mathbf{q}$, denoted $\dfrac{\delta F}{\delta \mathbf{q}}(\mathbf{q})$, is defined as the *Fréchet* derivative of the functional $F(\mathbf{q})$, i.e.,

$$\lim_{\varepsilon \to 0} \frac{\partial}{\partial \varepsilon} F(\mathbf{q}+\varepsilon\delta\mathbf{q}) = \left\langle \frac{\delta F}{\delta \mathbf{q}}, \delta\mathbf{q} \right\rangle. \tag{3.26}$$

Consider the following simple example. Let $I(u)$ be the functional

$$I(u) = \int_0^1 u^4 - u_x(u_{xx})^3 \, dx, \tag{3.27}$$

where $u = u(x)$ is a smooth scalar function satisfying the Dirichlet conditions $u(0) = u_0$ and $u(1) = u_1$. Note that the density, given by $u^4 - u_x(u_{xx})^3$, explicitly depends not only on u but also on u_x and u_{xx}.

The variational derivative $\dfrac{\delta I}{\delta u}$ is computed as follows. We have

$$I(u + \varepsilon\delta u) = \int_0^1 (u + \varepsilon\delta u)^4 - (u + \varepsilon\delta u)_x \left[(u + \varepsilon\delta u)_{xx}\right]^3 \, dx,$$

from which it follows that

$$\frac{\partial}{\partial \varepsilon}I(u + \varepsilon\delta u) = \int_0^1 \{4\delta u\,(u + \varepsilon\delta u)^3 - \delta u_x \left[(u + \varepsilon\delta u)_{xx}\right]^3$$

$$-3\delta u_{xx}\,(u + \varepsilon\delta u)_x \left[(u + \varepsilon\delta u)_{xx}\right]^2 \}dx,$$

and consequently that

$$\lim_{\varepsilon \to 0} \frac{\partial}{\partial \varepsilon}I(u + \varepsilon\delta u) = \int_0^1 4u^3\delta u - (u_{xx})^3\,\delta u_x - 3u_x\,(u_{xx})^2\,\delta u_{xx} \, dx.$$

To identify the variational derivative we must integrate this expression by parts until the integral can be put into the form

$$\int_0^1 \frac{\delta I}{\delta u} \delta u \, dx \equiv \left\langle \frac{\delta I}{\delta u}, \delta u \right\rangle,$$

as required by (3.26). The integration by parts requires boundary conditions on δu at $x = 0$ and $x = 1$, respectively. We impose the *natural* boundary conditions that δu and *all* of its derivatives are *zero* at $x = 0$ and $x = 1$, respectively. Consequently it follows that

$$\int_0^1 4u^3 \delta u - (u_{xx})^3 \delta u_x - 3u_x (u_{xx})^2 \delta u_{xx} \, dx$$

$$= \int_0^1 4u^3 \delta u + 3 (u_{xx})^2 u_{xxx} \delta u + 3 \left[u_x (u_{xx})^2 \right]_x \delta u_x \, dx$$

$$= \int_0^1 \left[4u^3 + 3 (u_{xx})^2 u_{xxx} - 3 \left(u_x (u_{xx})^2 \right)_{xx} \right] \delta u \, dx$$

$$= \left\langle 4u^3 + 3 (u_{xx})^2 u_{xxx} - 3 \left(u_x (u_{xx})^2 \right)_{xx}, \delta u \right\rangle.$$

Thus we conclude that

$$\frac{\delta I}{\delta u} = 4u^3 + 3 (u_{xx})^2 u_{xxx} - 3 \left(u_x (u_{xx})^2 \right)_{xx}. \tag{3.28}$$

In practice it is usually easiest to obtain the variational derivative $\dfrac{\delta F}{\delta \mathbf{q}}$ from the *first variation of F*, denoted by δF, as computed using the techniques of variational calculus (e.g., Gelfand & Fomin, 1963). Formally,

$$\delta F = \delta \int_\Omega \mathcal{F}(\mathbf{q}) d\mathbf{x} = \int_\Omega \delta \mathcal{F}(\mathbf{q}) d\mathbf{x}$$

$$= \int_\Omega \left(\frac{\delta F}{\delta \mathbf{q}} \right)^\top \delta \mathbf{q} \, d\mathbf{x} = \left\langle \frac{\delta F}{\delta \mathbf{q}}, \delta \mathbf{q} \right\rangle, \tag{3.29}$$

assuming that the boundary of Ω is fixed.

For example, the first variation of $I(u)$ is given by

$$\delta I = \int_0^1 \delta \left[u^4 - u_x (u_{xx})^3 \right] \, dx$$

$$= \int_0^1 \left[4u^3 \delta u - \delta u_x (u_{xx})^3 - 3u_x (u_{xx})^2 \delta u_{xx} \right] \, dx$$

$$= \int_0^1 \left[4u^3 + 3 (u_{xx})^2 u_{xxx} - 3 \left(u_x (u_{xx})^2 \right)_{xx} \right] \delta u \, dx, \tag{3.30}$$

after integrating by parts. It follows from (3.29) and (3.30) that

$$\frac{\delta I}{\delta u} = 4u^3 + 3 (u_{xx})^2 u_{xxx} - 3 \left(u_x (u_{xx})^2 \right)_{xx},$$

which is, of course, identical to (3.28).

Poisson bracket formulation of the dynamics

It is possible to rewrite the Hamiltonian system (3.18) using the Poisson bracket (3.19). Suppose that $F(\mathbf{q})$ is an arbitrary functional of $\mathbf{q}$ of the form

$$F(\mathbf{q}) = \int_\Omega \mathcal{F}(\mathbf{q}) d\mathbf{x}.$$

The partial derivative with respect to time of F is determined as follows

$$\frac{\partial F}{\partial t} = \frac{\partial}{\partial t} \int_\Omega \mathcal{F}(\mathbf{q}) d\mathbf{x} = \int_\Omega \frac{\partial \mathcal{F}}{\partial t} d\mathbf{x} = \int_\Omega \left(\frac{\delta F}{\delta \mathbf{q}} \right)^{\mathsf{T}} \mathbf{q}_t \, d\mathbf{x}$$

$$= \int_\Omega \left(\frac{\delta F}{\delta \mathbf{q}} \right)^{\mathsf{T}} \mathbf{J} \frac{\delta H}{\delta \mathbf{q}} \, d\mathbf{x} = \left\langle \frac{\delta F}{\delta \mathbf{q}}, \mathbf{J} \frac{\delta H}{\delta \mathbf{q}} \right\rangle,$$

where (3.18) has been used and where it has been assumed that Ω is fixed. Thus we conclude, using (3.19), that

$$\frac{\partial F}{\partial t} = [F, H]. \tag{3.31}$$

It therefore follows that

$$\frac{\partial \mathbf{q}}{\partial t} = [\mathbf{q}, H],\tag{3.32}$$

provided we interpret $\mathbf{q}(\mathbf{x}, t)$ as the *functional*

$$\mathbf{q}(\mathbf{x}, t) = \int_{\Omega} \mathbf{q}(\mathbf{x}', t)\, \delta(\mathbf{x}' - \mathbf{x})\, d\mathbf{x}',\tag{3.33}$$

where $\delta(\mathbf{x}' - \mathbf{x})$ is the *delta function* centered at $\mathbf{x}' = \mathbf{x}$.

To verify that (3.32) indeed reproduces (3.18) we need to compute the variational derivative of (3.33). The first variation is given by

$$\delta \mathbf{q}(\mathbf{x}, t) = \int_{\Omega} \delta \mathbf{q}(\mathbf{x}', t)\, \delta(\mathbf{x}' - \mathbf{x})\, d\mathbf{x}' = \langle \delta(\mathbf{x}' - \mathbf{x}), \delta \mathbf{q} \rangle,$$

from which we conclude

$$\frac{\delta \mathbf{q}}{\delta \mathbf{q}} = \delta(\mathbf{x}' - \mathbf{x}).$$

If this expression is substituted into (3.32) it follows that

$$\frac{\partial \mathbf{q}(\mathbf{x}, t)}{\partial t} = [\mathbf{q}, H]$$

$$= \left\langle \frac{\delta \mathbf{q}}{\delta \mathbf{q}}, \mathbf{J} \frac{\delta H}{\delta \mathbf{q}} \right\rangle = \int_{\Omega} \delta(\mathbf{x}' - \mathbf{x})\, \mathbf{J}(\mathbf{x}', t) \frac{\delta H}{\delta \mathbf{q}}(\mathbf{x}', t)\, d\mathbf{x}'$$

$$= \mathbf{J}(\mathbf{x}, t) \frac{\delta H}{\delta \mathbf{q}}(\mathbf{x}, t),\tag{3.34}$$

which is (3.18).

3.3   Hamiltonian structure of the Euler equations

In this section we will derive a noncanonical Hamiltonian formulation for the vorticity equation (3.8). The notation used here will be identical to that introduced in Section 3.1.

Theorem 3.1 *The vorticity equation (3.8) can be written as a scalar Hamiltonian system with*

$$q = \Delta\psi, \tag{3.35}$$

$$J(*) = -\partial(q, *), \tag{3.36}$$

$$H = \frac{1}{2}\iint_\Omega \nabla\psi \cdot \nabla\psi \, dxdy - \sum_{i=1}^n \lambda_i \oint_{\partial\Omega_i} \mathbf{n}_i \cdot \nabla\psi \, ds, \tag{3.37}$$

where $\mathbf{n}_i$ is the unit outward normal on the boundary curve $\partial\Omega_i$ for $i = 1, ..., n$.

Proof. The proof of this theorem requires several steps. First, we will show that (3.8) can be written in the form (3.18) for q, J and H given by (3.35), (3.36) and (3.37), respectively. Second, we will show that the Hamiltonian is an invariant of the motion. We will then show that the bracket (3.19) satisfies the five properties H1, H2, H3, H4 and H5.

To show that the vorticity equation can be written in the form (3.18), we must compute the variational derivative $\dfrac{\delta H}{\delta q}$. The first variation of the Hamiltonian H is given by

$$\delta H = \iint_\Omega \nabla\psi \cdot \nabla\delta\psi \, dxdy - \sum_{i=1}^n \lambda_i \oint_{\partial\Omega_i} \mathbf{n}_i \cdot \nabla\delta\psi \, ds$$

$$= \iint_\Omega (-\psi)\,\Delta\delta\psi \, dxdy + \sum_{i=1}^n [\psi - \lambda_i]_{\partial\Omega_i} \oint_{\partial\Omega_i} \mathbf{n}_i \cdot \nabla\delta\psi \, ds$$

$$= \iint_\Omega (-\psi)\,\Delta\delta\psi \, dxdy = \iint_\Omega (-\psi)\,\delta q \, dxdy, \tag{3.38}$$

where we have integrated by parts once and exploited the boundary condition (3.15). It therefore follows that

$$\frac{\delta H}{\delta q} = -\psi. \tag{3.39}$$

Substitution of (3.39) into (3.36) implies

$$J\left(q, \frac{\delta H}{\delta q}\right) = -\partial(q, \psi) = \partial(\psi, \Delta\psi) = \Delta\psi_t = q_t,$$

where (3.8) and property *J2* has been used. We have therefore shown that (3.8) can be written in the form (3.18)

We now show that H is an invariant of the motion. It follows that

$$\frac{dH}{dt} = \iint_\Omega \nabla\psi \cdot \nabla\psi_t \, dxdy - \sum_{i=1}^n \lambda_i \oint_{\partial\Omega_i} \mathbf{n}_i \cdot \nabla\psi_t \, ds$$

$$= -\iint_\Omega \psi\Delta\psi_t \, dxdy + \sum_{i=1}^n [\psi - \lambda_i]_{\partial\Omega_i} \oint_{\partial\Omega_i} \mathbf{n}_i \cdot \nabla\psi_t \, ds$$

$$= -\iint_\Omega \psi\Delta\psi_t \, dxdy = \iint_\Omega \psi\partial(\psi, \Delta\psi) \, dxdy$$

$$= \frac{1}{2}\iint_\Omega \partial(\psi^2, \Delta\psi) \, dxdy$$

$$= \frac{1}{2}\iint_\Omega \nabla \cdot \left[(\mathbf{e}_3 \times \nabla\psi^2)\,\Delta\psi\right] \, dxdy$$

$$= \frac{1}{2}\sum_{i=1}^n \oint_{\partial\Omega_i} \mathbf{n}_i \cdot (\mathbf{e}_3 \times \nabla\psi^2)\,\Delta\psi \, ds$$

$$= -\sum_{i=1}^n \oint_{\partial\Omega_i} (\mathbf{t}_i \cdot \nabla\psi)\psi\Delta\psi \, ds = 0,$$

since

$$[\mathbf{t}_i \cdot \nabla\psi]_{\partial\Omega_i} = \left[\frac{\partial\psi}{\partial s}\right]_{\partial\Omega_i} = 0,$$

for each $i = 1, ..., n$, where $\mathbf{t}_i$ is the unit tangent vector on $\partial\Omega_i$ and where we have integrated by parts once, used property *J4*, and exploited the boundary condition (3.15).

What remains to be shown is that the bracket

$$[F, G] \equiv \left\langle \frac{\delta F}{\delta q}, J\frac{\delta G}{\delta q} \right\rangle = -\iint_\Omega \frac{\delta F}{\delta q}\partial\left(q, \frac{\delta G}{\delta q}\right) \, dxdy, \tag{3.40}$$

where F and G are functionals of q, satisfy properties *H1, H2, H3, H4* and *H5* of the general definition.

Before we begin this demonstration it is necessary to remark on the class of allowed functionals. The proof of these five properties will require integrating by parts. In order that the boundary integrals vanish we must restrict the set of allowed functionals to those whose variational derivatives evaluated on the boundary can either be explicitly written as a function of q (i.e., as we shall see, the Casimirs) or those whose first variational derivatives are constant on each simply connected portion of the boundary.

That is, if F is an allowed functional, then either

$$\left[\frac{\delta F}{\delta q}\right]_{\partial \Omega_i} = f_i(q),$$

where $f_i(q)$ is a suitably smooth function of its argument, or

$$\left[\frac{\partial}{\partial s}\left(\frac{\delta F}{\delta q}\right)\right]_{\partial \Omega_i} = 0,$$

for each $i = 1, ..., n$. For example, $\dfrac{\delta H}{\delta q} = -\psi$ which is constant on $\partial \Omega_i$. Another example is the enstrophy functional, given by

$$\iint\limits_{\Omega} q^2 \, dx dy,$$

which gives

$$\frac{\delta}{\delta q} \iint\limits_{\Omega} q^2 \, dx dy = 2q,$$

so that, in this case, $f_i(q) = 2q$. A functional which satisfies this property will be called an *allowable functional*.

The first property we show is self-commutation, i.e., $[F, F] = 0$. It follows from (3.40) that

$$[F, F] = -\iint\limits_{\Omega} \frac{\delta F}{\delta q} \partial\left(q, \frac{\delta F}{\delta q}\right) dx dy$$

$$= -\frac{1}{2} \iint\limits_{\Omega} \nabla \cdot \left[\left(\frac{\delta F}{\delta q}\right)^2 \mathbf{e}_3 \times \nabla(q)\right] dx dy$$

$$= -\frac{1}{2} \sum_{i=1}^{n} \oint_{\partial \Omega_i} \left(\frac{\delta F}{\delta q} \right)^2 \mathbf{n} \cdot (\mathbf{e}_3 \times \nabla(q)) \, ds$$

$$= \frac{1}{2} \sum_{i=1}^{n} \oint_{\partial \Omega_i} \left(\frac{\delta F}{\delta q} \right)^2 \frac{\partial q}{\partial s} ds = \frac{1}{2} \sum_{i=1}^{n} \oint_{\partial \Omega_i} \chi_i(q) \frac{\partial q}{\partial s} ds,$$

$$= \frac{1}{2} \sum_{i=1}^{n} \oint_{\partial \Omega_i} \frac{\partial}{\partial s} \left[\int^{q} \chi_i(\xi) \, d\xi \right] \, ds = 0,$$

because of the periodicity around each closed curve $\partial \Omega_i$, and where for notational convenience we have introduced

$$\chi_i(q) \equiv \left[\left(\frac{\delta F}{\delta q} \right)^2 \right]_{\partial \Omega_i}$$

The second property we show is skew symmetry, i.e., $[F,G] = -[G,F]$. It follows from (3.40) that

$$[F,G] = - \iint_{\Omega} \frac{\delta F}{\delta q} \partial \left(q, \frac{\delta G}{\delta q} \right) dx dy$$

$$= - \iint_{\Omega} \frac{\delta F}{\delta q} \nabla \cdot \left[\frac{\delta G}{\delta q} \mathbf{e}_3 \times \nabla(q) \right] dx dy$$

$$= \iint_{\Omega} \frac{\delta G}{\delta q} \nabla \cdot \left[\frac{\delta F}{\delta q} \mathbf{e}_3 \times \nabla(q) \right] dx dy$$

$$- \sum_{i=1}^{n} \oint_{\partial \Omega_i} \frac{\delta G}{\delta q} \frac{\delta F}{\delta q} \mathbf{n} \cdot (\mathbf{e}_3 \times \nabla(q)) \, ds$$

$$= - [G,F] + \sum_{i=1}^{n} \oint_{\partial \Omega_i} \chi_i(q) \frac{\partial q}{\partial s} ds$$

$$= -[G, F] + \sum_{i=1}^{n} \oint_{\partial \Omega_i} \frac{\partial}{\partial s} \left[\int^q \chi_i(\xi) \, d\xi \right] \, ds = -[G, F],$$

where the boundary integrals are all zero because of the periodicity around each closed curve $\partial \Omega_i$, and where for notational convenience we have introduced

$$\chi_i(q) \equiv \left[\frac{\delta G}{\delta q} \frac{\delta F}{\delta q} \right]_{\partial \Omega_i}.$$

The third property we show is the distributive property, i.e., $[\alpha F + \beta G, Q] = \alpha \, [F, Q] + \beta \, [G, Q]$. It follows from (3.40) that

$$[\alpha F + \beta G, Q] = - \iint_\Omega \frac{\delta}{\delta q} (\alpha F + \beta G) \, \partial \left(q, \frac{\delta Q}{\delta q} \right) \, dx dy$$

$$= -\alpha \iint_\Omega \frac{\delta F}{\delta q} \partial \left(q, \frac{\delta Q}{\delta q} \right) \, dx dy - \beta \iint_\Omega \frac{\delta G}{\delta q} \partial \left(q, \frac{\delta Q}{\delta q} \right) \, dx dy$$

$$= \alpha \, [F, Q] + \beta \, [G, Q].$$

The fourth property we show is the associative property, i.e., $[FG, Q] = F \, [G, Q] + [F, Q] \, G$. It follows from (3.40) that

$$[FG, Q] = - \iint_\Omega \frac{\delta}{\delta q} (FG) \, \partial \left(q, \frac{\delta Q}{\delta q} \right) \, dx dy$$

$$= - \iint_\Omega \left(F \frac{\delta G}{\delta q} + G \frac{\delta F}{\delta q} \right) \partial \left(q, \frac{\delta Q}{\delta q} \right) \, dx dy$$

$$= -F \iint_\Omega \frac{\delta G}{\delta q} \partial \left(q, \frac{\delta Q}{\delta q} \right) \, dx dy - G \iint_\Omega \frac{\delta F}{\delta q} \partial \left(q, \frac{\delta Q}{\delta q} \right) \, dx dy$$

$$= F \, [G, Q] + [F, Q] \, G,$$

where the fact that F and G, as functionals, are functions of time only and can therefore be pulled outside the spatial integrals.

The last property we show is the Jacobi identity, i.e.,

$$[F, [G, Q]] + [G, [Q, F]] + [Q, [F, G]] \equiv 0,$$

for all allowable functionals F, G, and Q. The demonstration given here is similar to an argument presented by Scinocca & Shepherd (1992) for a similar problem. The first thing to observe is that if *any* of the functionals F, G, and Q are Casimirs, the Jacobi identity trivially holds, since by definition

$$[F, [G, Q]] = [G, [Q, F]] = [Q, [F, G]] \equiv 0,$$

in this situation.

It follows from the skew symmetry property and (3.40) that

$$[F, [G, Q]] + [G, [Q, F]] + [Q, [F, G]]$$

$$= -[[G, Q], F] - [[Q, F], G] - [[F, G], Q]$$

$$= -\left\langle \frac{\delta}{\delta q}[G, Q], J\frac{\delta F}{\delta q} \right\rangle - \left\langle \frac{\delta}{\delta q}[Q, F], J\frac{\delta G}{\delta q} \right\rangle$$

$$-\left\langle \frac{\delta}{\delta q}[F, G], J\frac{\delta Q}{\delta q} \right\rangle = -\left\langle \frac{\delta}{\delta q}\left\langle \frac{\delta G}{\delta q}, J\frac{\delta Q}{\delta q} \right\rangle, J\frac{\delta F}{\delta q} \right\rangle$$

$$-\left\langle \frac{\delta}{\delta q}\left\langle \frac{\delta Q}{\delta q}, J\frac{\delta F}{\delta q} \right\rangle, J\frac{\delta G}{\delta q} \right\rangle - \left\langle \frac{\delta}{\delta q}\left\langle \frac{\delta F}{\delta q}, J\frac{\delta G}{\delta q} \right\rangle, J\frac{\delta Q}{\delta q} \right\rangle$$

$$= -\left\langle \frac{\delta}{\delta q}\langle G', JQ' \rangle, JF' \right\rangle$$

$$-\left\langle \frac{\delta}{\delta q}\langle Q', JF' \rangle, JG' \right\rangle - \left\langle \frac{\delta}{\delta q}\langle F', JG' \rangle, JQ' \right\rangle, \tag{3.41}$$

where we have introduced

$$(*)' \equiv \frac{\delta(*)}{\delta q}.$$

To simplify (3.41) further we must compute the variational derivatives

$$\frac{\delta}{\delta q}\langle G', JQ' \rangle, \quad \frac{\delta}{\delta q}\langle Q', JF' \rangle \quad \text{and} \quad \frac{\delta}{\delta q}\langle F', JG' \rangle.$$

We will explicitly compute the first of these and the rest will follow in an obvious way. We begin with

$$\frac{\delta}{\delta q} \langle G', JQ' \rangle = \frac{\delta}{\delta q} \iint_\Omega \frac{\delta G}{\delta q} J \frac{\delta Q}{\delta q} \, dxdy$$

$$= -\frac{\delta}{\delta q} \iint_\Omega \frac{\delta G}{\delta q} \partial \left(q, \frac{\delta Q}{\delta q} \right) \, dxdy.$$

The variational derivative can be obtained from the first variation, i.e.,

$$\delta \iint_\Omega \frac{\delta G}{\delta q} \partial \left(q, \frac{\delta Q}{\delta q} \right) \, dxdy = \iint_\Omega \{ \frac{\delta^2 G}{\delta q^2} \delta q \partial \left(q, \frac{\delta Q}{\delta q} \right)$$

$$+ \frac{\delta G}{\delta q} \partial \left(\delta q, \frac{\delta Q}{\delta q} \right) + \frac{\delta G}{\delta q} \partial \left(q, \frac{\delta^2 Q}{\delta q^2} \delta q \right) \} \, dxdy$$

$$= \iint_\Omega \{ \frac{\delta^2 G}{\delta q^2} \delta q \partial \left(q, \frac{\delta Q}{\delta q} \right) - \frac{\delta G}{\delta q} \nabla \cdot \left[\delta q \mathbf{e}_3 \times \nabla \left(\frac{\delta Q}{\delta q} \right) \right]$$

$$+ \frac{\delta G}{\delta q} \nabla \cdot \left[\frac{\delta^2 Q}{\delta q^2} \delta q \mathbf{e}_3 \times \nabla (q) \right] \} \, dxdy$$

$$= \iint_\Omega \{ \frac{\delta^2 G}{\delta q^2} \partial \left(q, \frac{\delta Q}{\delta q} \right) + \nabla \cdot \left[\frac{\delta G}{\delta q} \mathbf{e}_3 \times \nabla \left(\frac{\delta Q}{\delta q} \right) \right]$$

$$- \frac{\delta^2 Q}{\delta q^2} \nabla \cdot \left[\frac{\delta G}{\delta q} \mathbf{e}_3 \times \nabla (q) \right] \} \delta q \, dxdy$$

$$= \iint_\Omega \{ -\frac{\delta^2 G}{\delta q^2} J \frac{\delta Q}{\delta q} - \partial \left(\frac{\delta G}{\delta q}, \frac{\delta Q}{\delta q} \right) + \frac{\delta^2 Q}{\delta q^2} J \frac{\delta G}{\delta q} \} \delta q \, dxdy,$$

from which we conclude

$$\frac{\delta}{\delta q} \langle G', JQ' \rangle = G'' JQ' + \partial (G', Q') - Q'' JG', \tag{3.42}$$

where

$$(*)'' \equiv \frac{\delta^2(*)}{\delta q^2}.$$

We note that the boundary integrals associated with the various integrations by parts required to derive (3.42) have all vanished as a consequence of the definition an allowable functional and where we have, for convenience, introduced the notation

$$\delta\left(\frac{\delta*}{\delta q}\right) = \frac{\delta^2 *}{\delta q^2}\delta q.$$

Similarly, it follows that

$$\frac{\delta}{\delta q}\langle Q', JF'\rangle = Q''JF' + \partial(Q',F') - F''JQ', \qquad (3.43)$$

$$\frac{\delta}{\delta q}\langle F', JG'\rangle = F''JG' + \partial(F',G') - G''JF'. \qquad (3.44)$$

Substitution of (3.42), (3.43) and (3.44) into (3.41) leads to

$$[F,[G,Q]] + [G,[Q,F]] + [Q,[F,G]]$$

$$= -\langle G''JQ', JF'\rangle + \langle G''JF', JQ'\rangle$$

$$- \langle Q''JF', JG'\rangle + \langle Q''JG', JF'\rangle$$

$$- \langle F''JG', JQ'\rangle + \langle F''JQ', JG'\rangle - \langle \partial(G',Q'), JF'\rangle$$

$$- \langle \partial(Q',F'), JG'\rangle - \langle \partial(F',G'), JQ'\rangle. \qquad (3.45)$$

However, since

$$\langle G''JQ', JF'\rangle = \iint_{\Omega} G''(JQ')(JF')\, dxdy = \langle G''JF', JQ'\rangle,$$

$$\langle Q''JF', JG'\rangle = \iint_{\Omega} Q''(JF')(JG')\, dxdy = \langle Q''JG', JF'\rangle,$$

$$\langle F'' JG' , JQ' \rangle = \iint\limits_{\Omega} F''(JG')(JQ') \, dxdy = \langle F'' JQ' , JG' \rangle,$$

it follows that (3.45) simplifies to

$$[F, [G, Q]] + [G, [Q, F]] + [Q, [F, G]]$$

$$= - \langle \partial (G',Q') , JF' \rangle - \langle \partial (Q',F') , JG' \rangle - \langle \partial (F',G') , JQ' \rangle$$

$$= \langle \partial (G',Q') , \partial (q, F') \rangle$$

$$+ \langle \partial (Q',F') , \partial (q, G') \rangle + \langle \partial (F',G') , \partial (q, Q') \rangle$$

$$= \iint\limits_{\Omega} \{ \partial (G',Q') \, \partial (q, F')$$

$$+ \partial (Q',F') \, \partial (q, G') + \partial (F',G') \, \partial (q, Q') \} \, dxdy$$

$$= - \iint\limits_{\Omega} \{ \partial (G',Q') \nabla \cdot [q\mathbf{e_3} \times \nabla(F')]$$

$$+ \partial (Q',F') \nabla \cdot [q\mathbf{e_3} \times \nabla(G')]$$

$$+ \partial (F',G') \nabla \cdot [q\mathbf{e_3} \times \nabla(Q')] \} \, dxdy$$

$$= \sum_{i=1}^{n} \oint\limits_{\partial \Omega_i} [\partial (G',Q') \, F_s' + \partial (Q',F') \, G_s' + \partial (F',G') \, Q_s'] \, q \, ds$$

$$+ \iint\limits_{\Omega} \{ \partial (\partial (G',Q') , F') + \partial (\partial (Q',F') , G')$$

$$+ \partial (\partial (F',G') , Q') \} q \, dxdy, \tag{3.46}$$

where

$$(*)'_s \equiv \frac{\partial(*)'}{\partial s}.$$

The integrand in the second term of (3.46) is identically zero because the Jacobian $\partial(*_1, *_2)$ satisfies property $J5$ (see Section 3.1) and the first term is identically zero because the terms in the square bracket in the integrand can be written in the form

$$[\partial(G',Q')F'_s + \partial(Q',F')G'_s + \partial(F',G')Q'_s]_{\partial\Omega_i}$$

$$= [(G'_n Q'_s - G'_s Q'_n)F'_s$$

$$+(Q'_n F'_s - Q'_s F'_n)G'_s$$

$$+(F'_n G'_s - F'_s G'_n)Q'_s]_{\partial\Omega_i} = 0,$$

where

$$(*)'_n \equiv \frac{\partial(*)'}{\partial n},$$

and where we have written the Jacobian in terms of the normal and tangential derivatives on the boundary. Thus we have shown that the Jacobi identity

$$[F,[G,Q]] + [G,[Q,F]] + [Q,[F,G]] = 0,$$

holds for the bracket (3.40) for arbitrary functionals F, G and Q. This completes the proof of Theorem (3.1). ∎

Remarks on the Hamiltonian formulation

The Hamiltonian H is the sum of the area-integrated kinetic energy

$$KE \equiv \frac{1}{2} \iint_\Omega \nabla\psi \cdot \nabla\psi \, dxdy, \tag{3.47}$$

and the weighted sum of the circulations

$$\Gamma_i \equiv \oint_{\partial\Omega_i} \mathbf{n}_i \cdot \nabla\psi \, ds, \quad i = 1, ..., n, \tag{3.48}$$

where the weight associated with each circulation integral is the negative of the constant value the stream function must take on the corresponding smooth simply connected portion of the boundary. Note that the Hamiltonian is *not*, in general, the energy and is indefinite in sign. However, if the domain is simply connected, then we may, without loss of generality, set

$$\psi = \lambda = 0 \text{ on } \partial\Omega,$$

(if Ω has a boundary) and thus the contribution to H from the circulations may be omitted. In this case the Hamiltonian *is* the energy and is positive definite. The circulations can also be omitted if $\Omega = \mathbb{R}^2$ and it is assumed that ψ has smooth compact support.

In our proof of Theorem 3.1 above we showed that the Hamiltonian is an invariant of the motion. In fact, it is straightforward to show that the kinetic energy KE *and* the circulations $\{\Gamma_i\}_{i=1}^n$ are *individually* invariants of the motion, i.e.,

$$\frac{d}{dt} \iint_\Omega \nabla\psi \cdot \nabla\psi \, dxdy = 0, \tag{3.49}$$

$$\frac{d}{dt} \oint_{\partial\Omega_i} \mathbf{n}_i \cdot \nabla\psi \, ds = 0, \tag{3.50}$$

with the latter being simply the *Kelvin circulation theorem*. We leave these demonstrations as an exercise for the reader.

Another point that should be made concerns the fact that the Hamiltonian (3.37) has been written in terms of the stream function and not the vorticity as might have been expected since we have chosen the vorticity as the Hamiltonian variable. We have chosen this representation for the Hamiltonian because it is more convenient for the arguments we present here. However, there are other representations. For example, if the kinetic energy term is integrated by parts once, the Hamiltonian can be written in the form

$$H = -\frac{1}{2} \iint_\Omega \psi\Delta\psi \, dxdy + \frac{1}{2}\sum_{i=1}^n \lambda_i \oint_{\partial\Omega_i} \mathbf{n}_i \cdot \nabla\psi \, ds,$$

and if $q = \Delta\psi$ is formally inverted, i.e., $\psi = \Delta^{-1}q$, then this expression can be formally re-written as

$$H = -\frac{1}{2} \iint_\Omega q\Delta^{-1}q \, dxdy + \frac{1}{2}\sum_{i=1}^n \oint_{\partial\Omega_i} (\Delta^{-1}q)\mathbf{n}_i \cdot \nabla(\Delta^{-1}q) \, ds, \tag{3.51}$$

which can be interpreted as the Hamiltonian written explicitly in terms of the Hamiltonian variable q. Regardless of which representation is used, we will formally consider $H = H(q)$.

It important to emphasize that one needs to explicitly check that a proposed Hamiltonian formulation in fact satisfies the required Hamiltonian properties *H1*, *H2*, *H3*, *H4* and *H5*. For example, suppose Ω is simply connected and $\psi = 0$ on $\partial\Omega$, Christiansen & Zabusky (1973) proposed that

$$q = \Delta\psi, \tag{3.52}$$

$$J(*) = -\partial(\psi, *), \tag{3.53}$$

$$H = \frac{1}{2}\iint_\Omega (\Delta\psi)^2, \tag{3.54}$$

was a Hamiltonian formulation of the vorticity equation (3.8). It is straightforward to verify that

$$\Delta\psi_t + \partial(\psi, \Delta\psi) = 0 \iff q_t = J\frac{\delta H}{\delta q},$$

since $\dfrac{\delta H}{\delta q} = q$ as required. However, it can be shown that the bracket

$$[F, G] \equiv \left\langle \frac{\delta F}{\delta q}, J\frac{\delta G}{\delta q} \right\rangle = -\iint_\Omega \frac{\delta F}{\delta q}\partial\left(\psi, \frac{\delta G}{\delta q}\right) dxdy, \tag{3.55}$$

does *not* satisfy the Jacobi identity *(H5)*. Thus this formulation for the vorticity equation is *not* a Hamiltonian description.

Poisson bracket formulation

It follows from (3.31) and (3.40) that if $F = F(q)$ is an arbitrary functional of the vorticity, then

$$\frac{\partial F}{\partial t} = [F, H] \equiv -\iint_\Omega \frac{\delta F}{\delta q}\partial\left(q, \frac{\delta H}{\delta q}\right) dxdy$$

$$= -\iint_\Omega \frac{\delta F}{\delta q}\partial(\psi, q)\, dxdy = \iint_\Omega \frac{\delta F}{\delta q}\frac{\partial q}{\partial t}\, dxdy.$$

In particular, the vorticity equation (3.8) can be written in the form

$$q_t = [q, H],$$

provided we interpret the vorticity as the functional

$$q(x, y, t) = \iint\limits_{\Omega'} q(x', y', t)\delta(x - x')\delta(y - y')\, dx'dy',$$

where Ω' is the domain Ω parametrized in terms of (x', y').

There is an alternate way of writing the Poisson bracket which is sometimes more convenient. From (3.40) we have

$$[F, G] \equiv -\iint\limits_{\Omega} \frac{\delta F}{\delta q}\partial\left(q, \frac{\delta G}{\delta q}\right) dxdy$$

$$= \iint\limits_{\Omega} \frac{\delta F}{\delta q}\nabla \cdot \left[q\mathbf{e}_3 \times \nabla\frac{\delta G}{\delta q}\right] dxdy$$

$$= -\sum_{i=1}^{n} \oint\limits_{\partial\Omega_i} q\frac{\delta F}{\delta q}\frac{\partial}{\partial s}\{\frac{\delta G}{\delta q}\}ds - \iint\limits_{\Omega} q\nabla \cdot \left[\frac{\delta F}{\delta q}\mathbf{e}_3 \times \nabla\frac{\delta G}{\delta q}\right] dxdy$$

$$= -\sum_{i=1}^{n} \oint\limits_{\partial\Omega_i} q\frac{\delta F}{\delta q}\frac{\partial}{\partial q}\{\frac{\delta G}{\delta q}\}\frac{\partial q}{\partial s}ds + \iint\limits_{\Omega} q\partial\left(\frac{\delta F}{\delta q}, \frac{\delta G}{\delta q}\right) dxdy,$$

$$= -\sum_{i=1}^{n} \oint\limits_{\partial\Omega_i} \frac{\partial}{\partial s}[\int^q \chi(\xi)\,d\xi]\, ds + \iint\limits_{\Omega} q\partial\left(\frac{\delta F}{\delta q}, \frac{\delta G}{\delta q}\right) dxdy,$$

where

$$\chi(q) \equiv \left[q\frac{\delta F}{\delta q}\frac{\partial}{\partial q}\{\frac{\delta G}{\delta q}\}\right]_{\partial\Omega_i}.$$

However, because of the required periodicity associated with each simply connected boundary curve $\partial\Omega_i$ it follows that the boundary integrals individually vanish. Thus, we conclude that an alternate representation for the Poisson bracket is

$$[F, G] = \iint\limits_{\Omega} q\partial\left(\frac{\delta F}{\delta q}, \frac{\delta G}{\delta q}\right) dxdy. \tag{3.56}$$

3.4   Reduction of the canonical Poisson bracket

In our presentation of the Hamiltonian formulation of the vorticity equation, we simply *guessed* the correct **J** operator and went on to show that it satisfied the requisite properties. The calculations required to verify properties *H1, H2, H3, H4* and *H5*, while relatively straightforward, are in general somewhat lengthy and tedious. A much better procedure, particularly for physically relevant applications which are sufficiently different from previously worked out Hamiltonian formulations, is to derive the Poisson bracket written in terms of the desired variables from a systematic reduction of the *canonical* Poisson bracket written in terms of the canonical Lagrangian positions and momenta. In this section we will present a derivation of (3.40) from via a systematic reduction of the canonical Poisson bracket for the two dimensional incompressible Euler equations.

The canonical (Lagrangian) positions and momenta (per unit mass) are given by, respectively,

$$x = x(a,b,t), \text{ with } a = x(a,b,0), \tag{3.57}$$

$$y = y(a,b,t), \text{ with } b = y(a,b,0), \tag{3.58}$$

$$u = u(a,b,t), \tag{3.59}$$

$$v = v(a,b,t), \tag{3.60}$$

where the position at $t = 0$ is given by $\mathbf{x}_{t=0} \equiv (x,y)_{t=0} = \mathbf{a} \equiv (a,b)$. Because the spatial domain is fixed and does not change with time, $\mathbf{a} \in \Omega$ *and* $\mathbf{x} \in \Omega$. Therefore, the canonical (Lagrangian) Poisson bracket for the two dimensional Euler equations may be written in the form (c.f. the continuous analogue of (2.26))

$$[F,G]_L \equiv \iint_\Omega \{\frac{\delta F}{\delta x}\frac{\delta G}{\delta u} - \frac{\delta F}{\delta u}\frac{\delta G}{\delta x} + \frac{\delta F}{\delta y}\frac{\delta G}{\delta v} - \frac{\delta F}{\delta v}\frac{\delta G}{\delta y}\} \, d\mathbf{a}, \tag{3.61}$$

where F and G are arbitrary functionals of the canonical variables, i.e.,

$$F(\mathbf{u}) \equiv \iint_\Omega \mathcal{F}(\mathbf{u}) \, d\mathbf{a}, \tag{3.62}$$

$$G(\mathbf{u}) \equiv \iint_\Omega \mathcal{G}(\mathbf{u}) \, d\mathbf{a}, \tag{3.63}$$

where

$$\mathbf{u} \equiv (u_1, u_2, u_3, u_4)^{\top}, \tag{3.64}$$

$$u_1 \equiv x\left(\mathbf{a}, t\right), \tag{3.65}$$

$$u_2 \equiv y\left(\mathbf{a}, t\right), \tag{3.66}$$

$$u_3 \equiv u\left(\mathbf{a}, t\right), \tag{3.67}$$

$$u_4 \equiv v\left(\mathbf{a}, t\right), \tag{3.68}$$

and where $d\mathbf{a} \equiv da\,db$.

We may re-write the canonical Poisson bracket (3.61) in the (general) symplectic form

$$[F, G]_L = \iint\limits_{\Omega} \left(\frac{\delta F}{\delta \mathbf{u}}\right)^{\top} \mathbf{J}_L \frac{\delta G}{\delta \mathbf{u}} da,$$

$$= \iint\limits_{\Omega} \frac{\delta F}{\delta u_i} J_{Lij} \frac{\delta G}{\delta u_j} da, \tag{3.69}$$

where, for the two dimensional Euler equations, the matrix $\mathbf{J}_L$ is given by

$$\mathbf{J}_L \equiv \begin{bmatrix} 0 & 0 & 1 & 0 \\ 0 & 0 & 0 & 1 \\ -1 & 0 & 0 & 0 \\ 0 & -1 & 0 & 0 \end{bmatrix}. \tag{3.70}$$

To proceed further it is necessary to determine the general transformation rule for Poisson brackets of continuous systems. The presentation given here, while targeted for the problem at hand, will be sufficiently general that it will be directly applicable to an arbitrary change of variables in a continuous Hamiltonian system. The approach taken here is similar to that described by Shepherd (1990). As a further example, Kuroda (1990) presents a related calculation for the nonhydrostatic primitive equations.

General transformation theory for a Poisson bracket of continuous fields

Consider the *general* change of *independent variables* given by the diffeomorphism $\mathbf{x} = \mathbf{x}(\mathbf{a})$ and the *general* change of *dependent variables* $\mathbf{u}(\mathbf{a},t) \rightarrow \mathbf{w}(\mathbf{x},t)$, where $\mathbf{w} = (w_1, ..., w_m)$. Let $F(\mathbf{u})$ be an arbitrary functional written with respect to $\mathbf{u}$ and $\mathbf{a}$ and let $\widetilde{F}(\mathbf{w})$ be the representation of F with respect to $\mathbf{w}$ and $\mathbf{x}$, i.e.,

$$F(\mathbf{u}) \equiv \iint_{\Omega} \mathcal{F}(\mathbf{u})\, d\mathbf{a},$$

$$\widetilde{F}(\mathbf{w}) \equiv \iint_{\Omega} \widetilde{\mathcal{F}}(\mathbf{w})\, d\mathbf{x}, \tag{3.71}$$

where $d\mathbf{a}$ and $d\mathbf{x}$ are the appropriate volume differentials with respect to the $\mathbf{a}$ and $\mathbf{x}$ variables, respectively, and Ω is the *fixed* spatial domain.

Lemma 3.2 *The variational derivative* $\dfrac{\delta F}{\delta \mathbf{u}}$ *is related to the variational derivative* $\dfrac{\delta \widetilde{F}}{\delta \mathbf{w}}$ *through the relation*

$$\frac{\delta F}{\delta u_i} = \iint_{\Omega} \frac{\delta \widetilde{F}}{\delta w_k} \frac{\delta w_k}{\delta u_i} d\mathbf{x},$$

for $i = 1, ..., n$.

   Proof. We proceed directly. By the definition of the first variation (c.f., (3.29)), it follows that

$$\delta \widetilde{F} = \iint_{\Omega} \frac{\delta \widetilde{F}}{\delta w_k} \delta w_k\, d\mathbf{x}, \tag{3.72}$$

$$\delta F = \iint_{\Omega} \frac{\delta F}{\delta u_i} \delta u_i\, d\mathbf{a}. \tag{3.73}$$

We now formally write $w_k(\mathbf{x}, t)$ as a *functional written in terms of* $\mathbf{u}(\mathbf{a}, t)$, i.e.,

$$w_k = \iint_{\Omega} \overline{w}_k(u_i)\, d\mathbf{a},$$

where $\overline{w}_k(u_i)$ is the appropriate density function. It follows, again by the definition of the first variation, that

$$\delta w_k = \iint\limits_\Omega \frac{\delta w_k}{\delta u_i}\, \delta u_i d\mathbf{a}, \tag{3.74}$$

where the variational derivative has to be properly interpreted because, as we shall see, it contains delta functions. Substitution of (3.74) into (3.72) yields

$$\delta \widetilde{F} = \iint\limits_\Omega \frac{\delta \widetilde{F}}{\delta w_k} \left\{ \iint\limits_\Omega \frac{\delta w_k}{\delta u_i}\, \delta u_i d\mathbf{a} \right\}\, d\mathbf{x}$$

$$= \iint\limits_\Omega \left\{ \iint\limits_\Omega \frac{\delta \widetilde{F}}{\delta w_k} \frac{\delta w_k}{\delta u_i}\, d\mathbf{x} \right\}\, \delta u_i d\mathbf{a}, \tag{3.75}$$

where the order of integration has been interchanged. Comparing (3.75) with (3.74) implies

$$\frac{\delta F}{\delta u_i} = \iint\limits_\Omega \frac{\delta \widetilde{F}}{\delta w_k} \frac{\delta w_k}{\delta u_i} d\mathbf{x}. \ \blacksquare \tag{3.76}$$

With the transformation rule (3.76) it is now straightforward to derive the following general transformation rule for the Poisson bracket.

Theorem 3.3 *Let F and G be arbitrary functionals written with respect to the variables $\mathbf{u}(\mathbf{a}, t)$, and let $\widetilde{F}$ and $\widetilde{G}$ be the corresponding functionals, respectively, written with respect to the variables $\mathbf{w}(\mathbf{x}, t)$ with $\mathbf{a} \in \Omega$ and $\mathbf{x} \in \Omega$. Let the Poisson bracket between F and G written in terms of the variables $\mathbf{u}(\mathbf{a}, t)$ and $\mathbf{w}(\mathbf{x}, t)$ be denoted by $[F, G]_U$ and $\left[\widetilde{F}, \widetilde{G}\right]_W$, respectively, then*

$$\left[\widetilde{F}, \widetilde{G}\right]_W = \iint\limits_\Omega \iint\limits_\Omega \frac{\delta \widetilde{F}(\mathbf{x}_1)}{\delta w_k} [w_k(\mathbf{x}_1), w_l(\mathbf{x}_2)]_U \frac{\delta \widetilde{G}(\mathbf{x}_2)}{\delta w_l}\, d\mathbf{x}_1 d\mathbf{x}_2.$$

Proof. The Poisson bracket $[F, G]_U$, written using index notation, is defined (c.f., (3.40)) as

$$[F, G]_U = \iint\limits_\Omega \frac{\delta F}{\delta u_i} J_{U_{ij}} \frac{\delta G}{\delta u_j} d\mathbf{a}. \tag{3.77}$$

However, it follows from the transformation rule (3.76) that

$$\frac{\delta F}{\delta u_i} = \iint_\Omega \frac{\delta \widetilde{F}(\mathbf{x}_1)}{\delta w_k} \frac{\delta w_k(\mathbf{x}_1)}{\delta u_i} \, d\mathbf{x}_1, \tag{3.78}$$

$$\frac{\delta G}{\delta u_j} = \iint_\Omega \frac{\delta \widetilde{G}(\mathbf{x}_2)}{\delta w_l} \frac{\delta w_l(\mathbf{x}_2)}{\delta u_j} \, d\mathbf{x}_2. \tag{3.79}$$

Substitution of (3.78) and (3.79) into (3.77) yields

$$[F,G]_U = \iint_\Omega \left\{ \iint_\Omega \frac{\delta \widetilde{F}(\mathbf{x}_1)}{\delta w_k} \frac{\delta w_k(\mathbf{x}_1)}{\delta u_i} \, d\mathbf{x}_1 \right\}$$

$$\times J_{U_{ij}} \left\{ \iint_\Omega \frac{\delta \widetilde{G}(\mathbf{x}_2)}{\delta w_l} \frac{\delta w_l(\mathbf{x}_2)}{\delta u_j} d\mathbf{x}_2 \right\} \, d\mathbf{a}$$

$$= \iint_\Omega \iint_\Omega \frac{\delta \widetilde{F}(\mathbf{x}_1)}{\delta w_k} \left\{ \iint_\Omega \frac{\delta w_k(\mathbf{x}_1)}{\delta u_i} J_{U_{ij}} \frac{\delta w_l(\mathbf{x}_2)}{\delta u_j} \, d\mathbf{a} \right\}$$

$$\times \frac{\delta \widetilde{G}(\mathbf{x}_2)}{\delta w_l} \, d\mathbf{x}_1 d\mathbf{x}_2$$

$$= \iint_\Omega \iint_\Omega \frac{\delta \widetilde{F}(\mathbf{x}_1)}{\delta w_k} [w_k(\mathbf{x}_1), w_l(\mathbf{x}_2)]_U \frac{\delta \widetilde{G}(\mathbf{x}_2)}{\delta w_l} \, d\mathbf{x}_1 d\mathbf{x}_2, \tag{3.80}$$

where the order of integration has been interchanged. However, (3.80) is simply the Poisson bracket between F and G *written with respect to the variables* $\mathbf{w}(\mathbf{x},t)$, i.e.,

$$\left[\widetilde{F}, \widetilde{G}\right]_w = \iint_\Omega \iint_\Omega \frac{\delta \widetilde{F}(\mathbf{x}_1)}{\delta w_k} [w_k(\mathbf{x}_1), w_l(\mathbf{x}_2)]_U$$

$$\times \frac{\delta \widetilde{G}(\mathbf{x}_2)}{\delta w_l} \, d\mathbf{x}_1 d\mathbf{x}_2. \ \blacksquare \tag{3.81}$$

Reduction of the canonical Euler Poisson bracket

We wish to derive the Poisson bracket (3.40) by applying the transformation rule (3.81) to the canonical Poisson bracket $[F, G]_L$. That is, we wish to show that

$$\left[\widetilde{F}, \widetilde{G}\right]_E \equiv \iint\limits_\Omega \iint\limits_\Omega \frac{\delta \widetilde{F}(\mathbf{x}_1)}{\delta q} \left[q(\mathbf{x}_1), q(\mathbf{x}_2)\right]_L \frac{\delta \widetilde{G}(\mathbf{x}_2)}{\delta q} \, d\mathbf{x}_1 d\mathbf{x}_2, \qquad (3.82)$$

where $\mathbf{x}_i \equiv (x_i, x_i)$ and $d\mathbf{x}_i \equiv dx_i dy_i$ for $i = 1, 2$ (with no summation implied) with $q \equiv v_x - u_y$, is *identically* the Poisson bracket (3.40).

To compute the right-hand-side of (3.82) it is necessary to compute $[q(\mathbf{x}_1), q(\mathbf{x}_2)]_L$ and this in turn (c.f. (3.69)) requires computing the variational derivatives

$$\frac{\delta q(\mathbf{x}_1)}{\delta \mathbf{u}} \text{ and } \frac{\delta q(\mathbf{x}_2)}{\delta \mathbf{u}},$$

where $\mathbf{u}$ are the canonical variables (3.64). In what follows we will work with $q(\mathbf{x}_1)$ since the calculation for $q(\mathbf{x}_2)$ is identical. First, we write $q(\mathbf{x}_1)$ as the functional

$$q(\mathbf{x}_1) = \iint\limits_\Omega q(\mathbf{x}) \delta (\mathbf{x} - \mathbf{x}_1) \, d\mathbf{x}. \qquad (3.83)$$

We now re-write $q(\mathbf{x})$ as follows

$$q(x, y) \equiv v_x - u_y = \frac{\partial (v, y)}{\partial (x, y)} - \frac{\partial (u, x)}{\partial (x, y)}$$

$$= \frac{\partial (a, b)}{\partial (x, y)} \left\{ \frac{\partial (v, y)}{\partial (a, b)} - \frac{\partial (u, x)}{\partial (a, b)} \right\}. \qquad (3.84)$$

However, since the flow is incompressible, i.e.,

$$\frac{\partial (a, b)}{\partial (x, y)} = \frac{\partial (x, y)}{\partial (a, b)} = 1, \qquad (3.85)$$

(3.84) reduces to

$$q(x, y) = \frac{\partial (v, y)}{\partial (a, b)} - \frac{\partial (u, x)}{\partial (a, b)}. \qquad (3.86)$$

Substitution of (3.84) into (3.83) implies

$$q(\mathbf{x}_1) = \iint\limits_\Omega \left\{ \frac{\partial (v, y)}{\partial (a, b)} - \frac{\partial (u, x)}{\partial (a, b)} \right\} \delta_1 da, \qquad (3.87)$$

where $da = dadb$ and $\delta_1 \equiv \delta(\mathbf{x} - \mathbf{x}_1)$, where it understood that the arguments of the delta function are written in terms of the canonical variables.

The variational derivative $\dfrac{\delta q(\mathbf{x}_1)}{\delta \mathbf{u}}$ can be obtained from the first variation $\delta q(\mathbf{x}_1)$, given by

$$\delta q(\mathbf{x}_1) = \iint_{\Omega} \left\{ \left[\frac{\partial(\delta v, y)}{\partial(a, b)} + \frac{\partial(v, \delta y)}{\partial(a, b)} - \frac{\partial(\delta u, x)}{\partial(a, b)} - \frac{\partial(u, \delta x)}{\partial(a, b)} \right] \delta_1 \right.$$

$$\left. + \left[\frac{\partial(v, y)}{\partial(a, b)} - \frac{\partial(u, x)}{\partial(a, b)} \right] \left[\frac{\partial \delta_1}{\partial x} \delta x + \frac{\partial \delta_1}{\partial y} \delta y \right] \right\} \, d\mathbf{a}, \qquad (3.88)$$

where the derivatives of the delta functions have to be properly interpreted in the sense of generalized functions (e.g., Lighthill, 1962). We must integrate the first term by parts. Observe that

$$\frac{\partial(A, B)}{\partial(a, b)} \equiv A_a B_b - A_b B_a$$

$$= \nabla_{\mathbf{a}} \cdot (B\mathbf{e}_3 \times \nabla_{\mathbf{a}} A) = -\nabla_{\mathbf{a}} \cdot (A\mathbf{e}_3 \times \nabla_{\mathbf{a}} B),$$

where

$$\nabla_{\mathbf{a}} \equiv \left(\frac{\partial}{\partial a}, \frac{\partial}{\partial b} \right),$$

and A and B are arbitrary functions. Thus we have, for example,

$$\iint_{\Omega} \frac{\partial(\delta v, y)}{\partial(a, b)} \delta_1 \, d\mathbf{a} = - \iint_{\Omega} \nabla_{\mathbf{a}} \cdot (\delta v \mathbf{e}_3 \times \nabla_{\mathbf{a}} y) \delta_1 \, d\mathbf{a}$$

$$= \iint_{\Omega} \nabla_{\mathbf{a}} \cdot (\delta_1 \mathbf{e}_3 \times \nabla_{\mathbf{a}} y) \delta v \, d\mathbf{a} = \iint_{\Omega} \frac{\partial(y, \delta_1)}{\partial(a, b)} \delta v \, d\mathbf{a}, \qquad (3.89)$$

where we have integrated by parts assuming that *all* variations vanish on the boundary $\partial\Omega$. Similarly, it follows that

$$\iint_{\Omega} \frac{\partial(\delta u, x)}{\partial(a, b)} \delta_1 \, d\mathbf{a} = \iint_{\Omega} \frac{\partial(x, \delta_1)}{\partial(a, b)} \delta u \, d\mathbf{a}, \qquad (3.90)$$

$$\iint_{\Omega} \frac{\partial(v, \delta y)}{\partial(a, b)} \delta_1 \, d\mathbf{a} = \iint_{\Omega} \frac{\partial(\delta_1, v)}{\partial(a, b)} \delta y \, d\mathbf{a}, \qquad (3.91)$$

$$\iint\limits_{\Omega} \frac{\partial(u,\delta x)}{\partial(a,b)} \delta_1 \, da = \iint\limits_{\Omega} \frac{\partial(\delta_1,u)}{\partial(a,b)} \delta x \, da. \tag{3.92}$$

Substitution of (3.89) through to (3.92) into (3.88) yields

$$\delta q(\mathbf{x}_1) = \iint\limits_{\Omega} \{ \frac{\partial(x,\delta_1)}{\partial(a,b)} \delta u + \frac{\partial(y,\delta_1)}{\partial(a,b)} \delta v$$

$$+ \left[\frac{\partial(\delta_1,u)}{\partial(a,b)} + q(\mathbf{x})\frac{\partial\delta_1}{\partial x} \right] \delta x$$

$$+ \left[\frac{\partial(\delta_1,v)}{\partial(a,b)} + q(\mathbf{x})\frac{\partial\delta_1}{\partial y} \right] \delta y\} \, da, \tag{3.93}$$

where (3.86) has been used. Thus, we conclude

$$\frac{\delta q(\mathbf{x}_1)}{\delta u} = \frac{\partial(x,\delta_1)}{\partial(a,b)}, \tag{3.94}$$

$$\frac{\delta q(\mathbf{x}_1)}{\delta v} = \frac{\partial(y,\delta_1)}{\partial(a,b)}, \tag{3.95}$$

$$\frac{\delta q(\mathbf{x}_1)}{\delta x} = \frac{\partial(\delta_1,u)}{\partial(a,b)} + q(\mathbf{x})\frac{\partial\delta_1}{\partial x}, \tag{3.96}$$

$$\frac{\delta q(\mathbf{x}_1)}{\delta y} = \frac{\partial(\delta_1,v)}{\partial(a,b)} + q(\mathbf{x})\frac{\partial\delta_1}{\partial y}. \tag{3.97}$$

The calculation for the variational derivative $\dfrac{\delta q(\mathbf{x}_2)}{\delta u}$ proceeds in a similar manner. If we introduce the functional

$$q(\mathbf{x}_2) = \iint\limits_{\Omega} q(\mathbf{x})\delta_2 d\mathbf{x},$$

where $\delta_2 \equiv \delta(\mathbf{x} - \mathbf{x}_2)$, it follows that

$$\frac{\delta q(\mathbf{x}_2)}{\delta u} = \frac{\partial(x,\delta_2)}{\partial(a,b)}, \tag{3.98}$$

$$\frac{\delta q(\mathbf{x_2})}{\delta v} = \frac{\partial (y, \delta_2)}{\partial (a, b)}, \tag{3.99}$$

$$\frac{\delta q(\mathbf{x_2})}{\delta x} = \frac{\partial (\delta_2, u)}{\partial (a, b)} + q(\mathbf{x})\frac{\partial \delta_2}{\partial x}, \tag{3.100}$$

$$\frac{\delta q(\mathbf{x_2})}{\delta y} = \frac{\partial (\delta_2, v)}{\partial (a, b)} + q(\mathbf{x})\frac{\partial \delta_2}{\partial y}. \tag{3.101}$$

With the variational derivatives $\dfrac{\delta q(\mathbf{x_1})}{\delta \mathbf{u}}$ and $\dfrac{\delta q(\mathbf{x_2})}{\delta \mathbf{u}}$ computed we can evaluate $[q(\mathbf{x_1}), q(\mathbf{x_2})]_L$ as follows. From (3.61) we have

$$[q(\mathbf{x_1}), q(\mathbf{x_2})]_L \equiv \iint_\Omega \{ \frac{\delta q(\mathbf{x_1})}{\delta x}\frac{\delta q(\mathbf{x_2})}{\delta u} - \frac{\delta q(\mathbf{x_1})}{\delta u}\frac{\delta q(\mathbf{x_2})}{\delta x}$$

$$+ \frac{\delta q(\mathbf{x_1})}{\delta y}\frac{\delta q(\mathbf{x_2})}{\delta v} - \frac{\delta q(\mathbf{x_1})}{\delta v}\frac{\delta q(\mathbf{x_2})}{\delta y}\} \, da$$

$$= \iint_\Omega \{ \left[\frac{\partial (\delta_1, u)}{\partial (a, b)} + q(\mathbf{x})\frac{\partial \delta_1}{\partial x}\right] \frac{\partial (x, \delta_2)}{\partial (a, b)}$$

$$- \left[\frac{\partial (\delta_2, u)}{\partial (a, b)} + q(\mathbf{x})\frac{\partial \delta_2}{\partial x}\right] \frac{\partial (x, \delta_1)}{\partial (a, b)}$$

$$+ \left[\frac{\partial (\delta_1, v)}{\partial (a, b)} + q(\mathbf{x})\frac{\partial \delta_1}{\partial y}\right] \frac{\partial (y, \delta_2)}{\partial (a, b)}$$

$$- \left[\frac{\partial (\delta_2, v)}{\partial (a, b)} + q(\mathbf{x})\frac{\partial \delta_2}{\partial y}\right] \frac{\partial (y, \delta_1)}{\partial (a, b)}\} \, da. \tag{3.102}$$

The integrand can be simplified by noting

$$\frac{\partial (A, B)}{\partial (a, b)} = \frac{\partial (A, B)}{\partial (x, y)}\frac{\partial (x, y)}{\partial (a, b)} = \frac{\partial (A, B)}{\partial (x, y)},$$

due to incompressibility, for arbitrary A and B, implying

$$\frac{\partial (x, \delta_2)}{\partial (a, b)} = \frac{\partial (x, \delta_2)}{\partial (x, y)} = \frac{\partial \delta_2}{\partial y},$$

$$\frac{\partial(x,\delta_1)}{\partial(a,b)} = \frac{\partial(x,\delta_1)}{\partial(x,y)} = \frac{\partial\delta_1}{\partial y},$$

$$\frac{\partial(y,\delta_2)}{\partial(a,b)} = \frac{\partial(y,\delta_2)}{\partial(x,y)} = -\frac{\partial\delta_2}{\partial x},$$

$$\frac{\partial(y,\delta_1)}{\partial(a,b)} = \frac{\partial(y,\delta_1)}{\partial(x,y)} = -\frac{\partial\delta_1}{\partial x}.$$

Substitution of these expressions into (3.102) leads to

$$[q(\mathbf{x}_1),q(\mathbf{x}_2)]_L \equiv \iint\limits_\Omega \left\{ 2q(\mathbf{x})\frac{\partial(\delta_1,\delta_2)}{\partial(x,y)} + \frac{\partial\delta_2}{\partial y}\frac{\partial(\delta_1,u)}{\partial(x,y)} \right.$$

$$\left. -\frac{\partial\delta_1}{\partial y}\frac{\partial(\delta_2,u)}{\partial(x,y)} - \frac{\partial\delta_2}{\partial x}\frac{\partial(\delta_1,v)}{\partial(x,y)} + \frac{\partial\delta_1}{\partial x}\frac{\partial(\delta_2,v)}{\partial(x,y)} \right\} d\mathbf{x}, \qquad (3.103)$$

where the integration is done with respect to the Eulerian variables $\mathbf{x}$, noting that the volume differential

$$d\mathbf{a} = \frac{\partial(a,b)}{\partial(x,y)}d\mathbf{x} = d\mathbf{x},$$

due to incompressibility. The integrand in (3.103) can be simplified still further by observing

$$\frac{\partial\delta_2}{\partial y}\frac{\partial(\delta_1,u)}{\partial(x,y)} - \frac{\partial\delta_1}{\partial y}\frac{\partial(\delta_2,u)}{\partial(x,y)} - \frac{\partial\delta_2}{\partial x}\frac{\partial(\delta_1,v)}{\partial(x,y)} + \frac{\partial\delta_1}{\partial x}\frac{\partial(\delta_2,v)}{\partial(x,y)}$$

$$= \frac{\partial\delta_2}{\partial y}\left[\frac{\partial\delta_1}{\partial x}\frac{\partial u}{\partial y} - \frac{\partial\delta_1}{\partial y}\frac{\partial u}{\partial x}\right] - \frac{\partial\delta_1}{\partial y}\left[\frac{\partial\delta_2}{\partial x}\frac{\partial u}{\partial y} - \frac{\partial\delta_2}{\partial y}\frac{\partial u}{\partial x}\right]$$

$$-\frac{\partial\delta_2}{\partial x}\left[\frac{\partial\delta_1}{\partial x}\frac{\partial v}{\partial y} - \frac{\partial\delta_1}{\partial y}\frac{\partial v}{\partial x}\right] + \frac{\partial\delta_1}{\partial x}\left[\frac{\partial\delta_2}{\partial x}\frac{\partial v}{\partial y} - \frac{\partial\delta_2}{\partial y}\frac{\partial v}{\partial x}\right]$$

$$= \left(\frac{\partial u}{\partial y} - \frac{\partial v}{\partial x}\right)\left(\frac{\partial\delta_1}{\partial x}\frac{\partial\delta_2}{\partial y} - \frac{\partial\delta_1}{\partial y}\frac{\partial\delta_2}{\partial x}\right) = -q(\mathbf{x})\frac{\partial(\delta_1,\delta_1)}{\partial(x,y)}.$$

Substitution of this expression into (3.103) implies

$$[q(\mathbf{x}_1),q(\mathbf{x}_2)]_L = \iint\limits_\Omega q(\mathbf{x})\frac{\partial(\delta_1,\delta_2)}{\partial(x,y)}d\mathbf{x}. \qquad (3.104)$$

Substitution (3.104) into (3.82) implies

$$\left[\widetilde{F},\widetilde{G}\right]_E \equiv \iint\limits_{\Omega} \iint\limits_{\Omega} \frac{\delta \widetilde{F}\left(\mathbf{x}_1\right)}{\delta q}$$

$$\times \left\{ \iint\limits_{\Omega} q(\mathbf{x}) \frac{\partial\left(\delta_1, \delta_2\right)}{\partial\left(x, y\right)}\, d\mathbf{x} \right\} \frac{\delta \widetilde{G}\left(\mathbf{x}_2\right)}{\delta q}\, d\mathbf{x}_1 d\mathbf{x}_2, \qquad (3.105)$$

for arbitrary functionals $\widetilde{F}$ and $\widetilde{G}$. This expression can be integrated by parts as follows. First, observe that

$$\iint\limits_{\Omega} q(\mathbf{x}) \frac{\partial\left(\delta_1, \delta_2\right)}{\partial\left(x, y\right)}\, d\mathbf{x} = \iint\limits_{\Omega} q\nabla \cdot \left[\delta_2 \mathbf{e}_3 \times \nabla \delta_1\right]\, d\mathbf{x}$$

$$= - \iint\limits_{\Omega} \delta_2 \nabla \cdot \left[q \mathbf{e}_3 \times \nabla \delta_1\right]\, d\mathbf{x}$$

$$\iint\limits_{\Omega} \delta_2 \frac{\partial\left(q, \delta_1\right)}{\partial\left(x, y\right)}\, d\mathbf{x} = \left[\frac{\partial\left(q, \delta_1\right)}{\partial\left(x, y\right)}\right]_{\mathbf{x}=\mathbf{x}_2} = \frac{\partial\left(\widehat{q}, \widehat{\delta}_1\right)}{\partial\left(x_2, y_2\right)},$$

where $\widehat{q} = \widehat{q}(\mathbf{x}_2) \equiv q(\mathbf{x}_2)$ and $\widehat{\delta}_1 \equiv \delta(\mathbf{x}_2 - \mathbf{x}_1)$. Substitution of this expression into (3.105) implies

$$\left[\widetilde{F},\widetilde{G}\right]_E = \iint\limits_{\Omega} \left\{ \iint\limits_{\Omega} \frac{\delta \widetilde{G}\left(\mathbf{x}_2\right)}{\delta q} \frac{\partial\left(\widehat{q}, \widehat{\delta}_1\right)}{\partial\left(x_2, y_2\right)} d\mathbf{x}_2 \right\} \frac{\delta \widetilde{F}\left(\mathbf{x}_1\right)}{\delta q}\, d\mathbf{x}_1$$

$$= \iint\limits_{\Omega} \left\{ \iint\limits_{\Omega} \frac{\delta \widetilde{G}\left(\mathbf{x}_2\right)}{\delta q} \nabla_{\mathbf{x}_2} \cdot \left[\widehat{\delta}_1 \mathbf{e}_3 \times \nabla_{\mathbf{x}_2} \widehat{q}\right] d\mathbf{x}_2 \right\} \frac{\delta \widetilde{F}\left(\mathbf{x}_1\right)}{\delta q} d\mathbf{x}_1$$

$$= - \iint\limits_{\Omega} \left\{ \iint\limits_{\Omega} \widehat{\delta}_1 \nabla_{\mathbf{x}_2} \cdot \left[\frac{\delta \widetilde{G}\left(\mathbf{x}_2\right)}{\delta q} \mathbf{e}_3 \times \nabla_{\mathbf{x}_2} \widehat{q}\right] d\mathbf{x}_2 \right\} \frac{\delta \widetilde{F}\left(\mathbf{x}_1\right)}{\delta q} d\mathbf{x}_1$$

$$
= \iint_{\Omega} \left\{ \iint_{\Omega} \hat{\delta}_1 \frac{\partial \left(\frac{\delta \tilde{G}(\mathbf{x}_2)}{\delta q}, \hat{q} \right)}{\partial (x_2, y_2)} d\mathbf{x}_2 \right\} \frac{\delta \tilde{F}(\mathbf{x}_1)}{\delta q} d\mathbf{x}_1
$$

$$
= \iint_{\Omega} \frac{\delta \tilde{F}(\mathbf{x}_1)}{\delta q} \left[\frac{\partial \left(\frac{\delta \tilde{G}(\mathbf{x}_2)}{\delta q}, \hat{q} \right)}{\partial (x_2, y_2)} \right]_{\mathbf{x}_2 = \mathbf{x}_1} d\mathbf{x}_1
$$

$$
= \iint_{\Omega} \frac{\delta \tilde{F}(\mathbf{x}_1)}{\delta q} \frac{\partial \left(\frac{\delta \tilde{G}(\mathbf{x}_1)}{\delta q}, \tilde{q}(\mathbf{x}_1) \right)}{\partial (x_1, y_1)} d\mathbf{x}_1, \tag{3.106}
$$

where $\tilde{q}(\mathbf{x}_1) \equiv \hat{q}(\mathbf{x}_1) \equiv q(\mathbf{x}_1)$. And, finally, this last expression can be re-written in the form

$$
\left[\tilde{F}, \tilde{G} \right]_E = \iint_{\Omega} \frac{\delta \tilde{F}(\mathbf{x})}{\delta q} \frac{\partial \left(\frac{\delta \tilde{G}(\mathbf{x})}{\delta q}, q(\mathbf{x}) \right)}{\partial (x, y)} d\mathbf{x}
$$

$$
= - \iint_{\Omega} \frac{\delta \tilde{F}}{\delta q} \frac{\partial \left(q, \frac{\delta \tilde{G}}{\delta q} \right)}{\partial (x, y)} d\mathbf{x} \equiv \left\langle \frac{\delta \tilde{F}}{\delta q}, J \frac{\delta \tilde{G}}{\delta q} \right\rangle,
$$

which is exactly the Poisson bracket (3.40) written in terms of the arbitrary functionals $\tilde{F}$ and $\tilde{G}$.

3.5   Casimir functionals

The definition for a *Casimir functional* is the obvious infinite dimensional analogue of the definition given for a Casimir function in Section 2.6 for a finite dimensional Hamiltonian system. A Casimir is a functional which Poisson commutes with every other functional. That is, given the general Hamiltonian system of partial differential equations (3.18) with Poisson bracket (3.19), the Casimirs are those functionals, denoted $C = C(\mathbf{q})$, satisfying

$$
[F, C] \equiv \left\langle \frac{\delta F}{\delta \mathbf{q}}, J \frac{\delta C}{\delta \mathbf{q}} \right\rangle = 0, \ \forall \ F(\mathbf{q}). \tag{3.107}
$$

Alternatively, we may describe Casimirs as that set of functionals which belong to the kernel of the Poisson bracket.

It follows from the properties of the inner product, that (3.107) implies that the complete set of Casimirs are the solutions of

$$\mathbf{J}\frac{\delta C}{\delta \mathbf{q}} = \mathbf{0}. \tag{3.108}$$

Thus, if $\mathbf{J}^{-1}$ exists, then

$$\frac{\delta C}{\delta \mathbf{q}} = \mathbf{0}, \tag{3.109}$$

which implies that the Casimirs are just the constant functionals with respect to the dependent variables. In this situation the Casimirs are said to be trivial and the Hamiltonian formulation is called *invertible*. However, if $\mathbf{J}^{-1}$ *does not exist*, then

$$\frac{\delta C}{\delta \mathbf{q}} \in \ker(\mathbf{J}), \tag{3.110}$$

which is nontrivial. If there exists nontrivial Casimirs, the Hamiltonian formulation is called *noninvertible*.

The Hamiltonian formulation for the vorticity equation (3.8) given in Theorem 3.1 is noninvertible because it has nontrivial Casimirs associated with. Substituting (3.35) and (3.36) into (3.108) implies that the Casimirs $C = C(q)$ are the solutions of

$$\partial\left(q, \frac{\delta C}{\delta q}\right) \equiv \frac{\partial\left(q, \frac{\delta C}{\delta q}\right)}{\partial(x, y)} = 0, \tag{3.111}$$

which implies that

$$\frac{\delta C}{\delta q} = \mathcal{C}(q), \tag{3.112}$$

where $\mathcal{C}(q)$ is an arbitrary *function* of q. It follows from (3.112) that the complete family of Casimir *functionals* for the Hamiltonian formulation of the vorticity equation may be written in the form

$$C(q) = \iint\limits_{\Omega} \left[\int^{q} \mathcal{C}(\xi)\,d\xi\right]\,dx dy. \tag{3.113}$$

Observe that

$$\delta C = \iint\limits_{\Omega} \frac{\partial}{\partial q}\left[\int^{q} \mathcal{C}(\xi)\,d\xi\right]\delta q\,dx dy$$

$$= \iint\limits_{\Omega} C(q)\delta q \; dxdy, \qquad (3.114)$$

which implies that

$$\frac{\delta C}{\delta q} = C\left(q\right),$$

which is (3.112). The Casimirs associated with the vorticity equation are, therefore, simply arbitrary integrals of the vorticity.

Since Casimirs Poisson commute with every other functional, it follows that they are constants of the motion since they must Poisson commute with the Hamiltonian, i.e.,

$$\frac{dC}{dt} = 0 \iff [H, C] = 0.$$

We can, however, also directly show the invariance of the Casimirs. From (3.113) we have

$$\frac{dC}{dt} = \iint\limits_{\Omega} \frac{\partial}{\partial q} \left[\int\limits^{q} C\left(\xi\right) d\xi \right] q_t \; dxdy$$

$$= \iint\limits_{\Omega} C(q) q_t \; dxdy = - \iint\limits_{\Omega} C(q)\partial\left(\psi, q\right) \; dxdy$$

$$= - \iint\limits_{\Omega} \partial\left(\psi, \left[\int\limits^{q} C\left(\xi\right) d\xi\right]\right) \; dxdy$$

$$= - \iint\limits_{\Omega} \nabla \cdot \left(\left[\int\limits^{q} C\left(\xi\right) d\xi\right] \mathbf{e_3} \times \nabla\psi\right) \; dxdy$$

$$= - \sum_{i=1}^{n} \oint\limits_{\partial\Omega_i} \mathbf{n}_i \cdot (\mathbf{e_3} \times \nabla\psi) \left[\int\limits^{q} C\left(\xi\right) d\xi\right] \; ds$$

$$= \sum_{i=1}^{n} \oint\limits_{\partial\Omega_i} \frac{\partial\psi}{\partial s} \left[\int\limits^{q} C\left(\xi\right) d\xi\right] \; ds = 0, \qquad (3.115)$$

where we have explicitly used (3.8) and the boundary conditions (3.15).

Even though Casimirs are invariants of the motion they *do not* necessarily correspond to a symmetry or invariance property of the Hamiltonian such as a translational or rotational invariance of the coordinates. Thus Casimirs *cannot*, in general, be obtained by applying Noether's Theorem (to be described in the next section). Casimirs, as we have shown, arise due to the noninvertibility or degeneracy of the J operator of the Hamiltonian formulation. As we shall see in Chapter 4, Casimirs are a crucial ingredient in establishing a variational principle for steady solutions to the vorticity equation (3.8).

3.6   Noether's Theorem

The connection between symmetry or invariance properties of the Hamiltonian structure and a particular conserved functional is expressed through Noether's Theorem. For the purposes of our discussion here we will introduce this theorem through an examination of the space-time translational invariance of the Hamiltonian formulation of the vorticity equation (3.8). For complete account of these ideas see, for example, Courant & Hilbert (1962) or Gelfand & Fomin (1963). Our presentation will sufficiently general that the extension to vector Hamiltonian systems will be straightforward.

We begin our discussion by observing that since the Hamiltonian H and operator J given, respectively, by (3.37) and (3.36) do not have an *explicit* dependence on any of the variables (x, y, t), they are obviously invariant under arbitrary translations in any one of the variables (x, y, t). To establish the connection between these invariances and an underlying conserved functional we need the following lemma

Lemma 3.4 *If H is invariant under arbitrary translations in the variable ξ, where ξ is any of (x, y, t), then*

$$\left\langle \frac{\delta H}{\delta q}, q_\xi \right\rangle = 0. \tag{3.116}$$

Proof. We will work with t since all the others follow similarly. Define the increment ΔH by

$$\Delta H \equiv H\left[q\left(x, y, t + \tau\right)\right] - H\left[q\left(x, y, t\right)\right]. \tag{3.117}$$

Since H is invariant under arbitrary translations, it follows that $\Delta H = 0 \; \forall \; \tau$ and thus *independently* of τ. If the Hamiltonian (3.37) is substituted into (3.117) and the result Taylor expanded about $\tau = 0$, we obtain

$$\Delta H = \tau \left[\iint_\Omega \nabla \psi \cdot \nabla \psi_t \, dx dy - \sum_{i=1}^n \lambda_i \oint_{\partial \Omega_i} \mathbf{n}_i \cdot \nabla \psi_t \, ds \right] + O\left(\tau^2\right). \tag{3.118}$$

The first integral can be integrated by parts once to give

$$\Delta H = -\tau \left[\iint_\Omega \psi q_t \, dxdy \right] + O\left(\tau^2\right). \tag{3.119}$$

The integral in the square brackets represents the principal linear part of the increment ΔH, and hence the variation in H is given by

$$\delta H = - \iint_\Omega \psi q_t \, dxdy$$

$$= \iint_\Omega \frac{\delta H}{\delta q} q_t \, dxdy = \left\langle \frac{\delta H}{\delta q}, q_t \right\rangle. \tag{3.120}$$

Finally, the fact that $\Delta H = 0 \ \forall \ \tau$ and thus *independently* of τ implies $\delta H = 0$ which by (3.120) is equivalent to

$$\left\langle \frac{\delta H}{\delta q}, q_t \right\rangle = 0. \ \blacksquare$$

Theorem 3.5 *(Noether's) Suppose H and J are invariant under arbitrary translations in the variable ξ, where ξ is any of (x, y, t) and let $M(q)$ be a functional satisfying*

$$J \frac{\delta M}{\delta q} = -q_\xi, \tag{3.121}$$

then M is invariant in time, i.e.,

$$\frac{dM}{dt} = 0. \tag{3.122}$$

Proof. We have

$$\frac{dM}{dt} \equiv [M, H] = -[H, M],$$

by the skew-symmetry property. However, substituting in the definition of the Poisson bracket we obtain

$$\frac{dM}{dt} = -\left\langle \frac{\delta H}{\delta q}, J \frac{\delta M}{\delta q} \right\rangle = \left\langle \frac{\delta H}{\delta q}, q_\xi \right\rangle = 0,$$

where (3.116) has been used. $\blacksquare$

Implicit in the proof of this theorem is the *existence* of a functional M satisfying (3.121). If J explicitly depends on the variable ξ then, in general, it is not possible to find such an M. This is the reason the statement of the theorem includes the requirement that J be invariant under translations in ξ. Another point to make is that the functional M must be an *allowable functional* in the sense defined in the proof of Theorem 3.1, i.e., the variational derivative $\frac{\delta M}{\delta q}$ evaluated on the boundary $\partial \Omega$ must be of the form $f(q)$.

Translational invariance with respect to x

As an example, let us determine the conserved functional corresponding to that fact that H and J do not explicitly depend on x. To make our discussion concrete the domain will be given by the periodic channel

$$\Omega = \{(x,y) \mid |x| < x_o, \ L_1 < y < L_2\},$$

with the boundary conditions $\psi = \lambda_1$ and λ_2 on $y = L_1$ and L_2, respectively, and where it is assumed that ψ is smoothly periodic at $x = \pm x_o$. Under these conditions, the Hamiltonian H given generally by (3.37), can be written in the reduced form

$$H = \frac{1}{2} \iint_\Omega \nabla\psi \cdot \nabla\psi \, dxdy + \lambda_2 \int_{-x_o}^{x_o} [u]_{y=L_2} \, dx$$

$$-\lambda_1 \int_{-x_o}^{x_o} [u]_{y=L_1} \, dx.$$

It is straightforward to check that introducing the transformation $x \to x + \tau$ leaves H and $J(*) \equiv -\partial(q, *)$ invariant for all τ. It therefore follows from Noether's Theorem that the functional M defined by

$$J\frac{\delta M}{\delta q} = -q_x, \tag{3.123}$$

is a constant of the motion. Substituting in for J we find

$$\partial\left(q, \frac{\delta M}{\delta q}\right) = q_x,$$

Hence we conclude that (modulo a Casimir)

$$\frac{\delta M}{\delta q} = y,$$

which implies that

$$M = \iint_\Omega yq \, dxdy. \tag{3.124}$$

The functional M is the x-component of *Kelvin's Impulse* (Lamb, 1932; see also Benjamin, 1984). Its relationship with the area-integrated x-component of linear momentum (per unit volume), given by,

$$\iint_\Omega u \, dxdy, \tag{3.125}$$

may be written in the form

$$M = \oint_{\partial\Omega} y\mathbf{n}\cdot\nabla\psi ds + \iint_{\Omega} u\; dxdy, \tag{3.126}$$

which is obtained by integrating (3.124) by parts once. The boundary integral in (3.126) can be further simplified to

$$\oint_{\partial\Omega} y\mathbf{n}\cdot\nabla\psi ds = L_2 \int_{-x_o}^{x_o} \psi_y\,(x,L_2,t)\; dx - L_1 \int_{-x_o}^{x_o} \psi_y\,(x,L_1,t)\; dx$$

$$= L_1 \int_{-x_o}^{x_o} u\,(x,L_1,t)\; dx - L_2 \int_{-x_o}^{x_o} u\,(x,L_2,t)\; dx,$$

where the contributions associated with the integrals on $x = \pm x_o$ cancel each other because of the assumed periodicity; specifically that

$$\psi_x\,(-x_o,y,t) = \psi_x\,(x_o,y,t)\,.$$

Kelvin's circulation theorem implies that

$$\frac{d}{dt} \int_{-x_o}^{x_o} u\,(x,L_1,t)\; dx = \frac{d}{dt} \int_{-x_o}^{x_o} u\,(x,L_1,t)\; dx = 0,$$

from which it follows that

$$\frac{d}{dt} \oint_{\partial\Omega} y\mathbf{n}\cdot\nabla\psi ds = 0. \tag{3.127}$$

This result, together with (3.126), implies that

$$\frac{dM}{dt} = 0 \iff \frac{d}{dt} \iint_{\Omega} u\; dxdy = 0.$$

Consequently, we have proved

Theorem 3.6 *The invariance of the Hamiltonian structure for the two dimensional vorticity equation to arbitrary translations in x implies that the x-direction linear momentum functional*

$$\iint_{\Omega} u\; dxdy,$$

is conserved by the dynamics.

A similar result can be established for the translational invariance with respect to y of the Hamiltonian structure. Assuming a periodic channel domain oriented along the y-axis (with corresponding boundary and periodicity conditions), it is left as an exercise for the reader to show that the invariance of the Hamiltonian structure for the two dimensional vorticity equation to arbitrary translations in y implies that the y-direction linear momentum functional

$$\iint_{\Omega} v \, dx dy, \tag{3.128}$$

is conserved by the dynamics.

Translational invariance with respect to t

Here we show that the invariance of the Hamiltonian structure to arbitrary translations in time implies that energy is conserved. Although energy conservation is already apparent from the fact that the Hamiltonian is a constant of the motion, it is useful to prove it as a consequence of Noether's Theorem. For this demonstration we will consider an arbitrary domain Ω as described in Section 3.1. The Hamiltonian structure is clearly invariant under arbitrary time translations since the domain is fixed in time and H and J have no explicit time dependence. It therefore follows from Noether's Theorem that the functional M satisfying

$$J\frac{\delta M}{\delta q} = -\partial\left(q, \frac{\delta M}{\delta q}\right) = -q_t,$$

is a constant of the motion. Comparing this expression with the vorticity equation (3.8) we conclude that (module a Casimir)

$$\frac{\delta M}{\delta q} = \psi.$$

However since

$$\frac{\delta H}{\delta q} = -\psi,$$

it follows that

$$M = -H,$$

where H is the Hamiltonian (3.37).

Thus

$$\frac{dM}{dt} = 0 \iff \frac{dH}{dt} = 0.$$

The time derivative of the all the circulation integrals in the Hamiltonian (3.37) are all identically zero because of Kelvin's circulation theorem. Hence we conclude, on the basis of Noether's Theorem that

$$\frac{d}{dt} \iint_\Omega \nabla\psi \cdot \nabla\psi \, dxdy = 0.$$

We have therefore proved

Theorem 3.7 *The invariance of the Hamiltonian structure for the two dimensional vorticity equation to arbitrary translations in t implies that the (kinetic) energy functional*

$$\iint_\Omega \nabla\psi \cdot \nabla\psi \, dxdy, \tag{3.129}$$

is conserved by the dynamics.

Invariance with respect to rotations

As a final example, we will show that the invariance of the Hamiltonian structure to rotations implies that the angular momentum is conserved. Consider the annulus domain given by

$$\Omega = \{(r, \theta) \mid 0 < a_1 < r < a_2, \ 0 \le \theta < 2\pi\},$$

with the boundary conditions $\psi = \lambda_1$ and λ_2 on $r = a_1$ and a_2, respectively, and where it is assumed that ψ is smoothly periodic at $\theta = 0$ and 2π. Here we have introduced the polar coordinates

$$x = r\cos(\theta), \quad y = r\sin(\theta). \tag{3.130}$$

Under these conditions, the Hamiltonian H given generally by (3.37), can be written in the reduced form

$$H = \frac{1}{2} \iint_\Omega \nabla\psi \cdot \nabla\psi \, rdrd\theta - \lambda_2 a_2 \int_0^{2\pi} [\psi_r]_{r=a_2} \, d\theta$$

$$+ \lambda_1 a_1 \int_0^{2\pi} [\psi_r]_{r=a_1} \, d\theta.$$

In the polar coordinates (3.130), the operator J takes the form

$$J(*) = -\frac{1}{r} [q_r (*)_\theta - (*)_r q_\theta]. \tag{3.131}$$

It is straightforward to check that introducing the transformation $\theta \to \theta + \tau$ leaves H and J $(*)$ invariant for all τ. It therefore follows from Noether's Theorem that the functional M defined by

$$J\frac{\delta M}{\delta q} = -q_\theta, \tag{3.132}$$

is a constant of the motion. Substitution of (3.131) into (3.132) yields

$$q_r \left(\frac{\delta M}{\delta q}\right)_\theta - \left(\frac{\delta M}{\delta q}\right)_r q_\theta = r q_\theta,$$

the solution of which can be written (modulo a Casimir)

$$\frac{\delta M}{\delta q} = -\frac{r^2}{2}. \tag{3.133}$$

Therefore the conserved functional associated with translation invariance with respect to θ is given by

$$M = -\frac{1}{2} \iint_\Omega r^2 q \, dxdy. \tag{3.134}$$

Its relationship with the area-integrated angular momentum (per unit volume), given by,

$$\iint_\Omega xv - yu \, dxdy, \tag{3.135}$$

may be written in the form

$$M = -\frac{1}{2} \sum_{i=1}^{2} \oint_{\partial\Omega_i} r^2 \mathbf{n}\cdot\nabla\psi ds + \iint_\Omega xv - yu \, dxdy, \tag{3.136}$$

which is obtained by integrating (3.134) by parts once. The boundary integral in (3.136) can be further simplified to

$$\sum_{i=1}^{2} \oint_{\partial\Omega_i} r^2 \mathbf{n}\cdot\nabla\psi ds$$

$$= (a_2)^3 \int_0^{2\pi} \psi_r (a_2,\theta,t) \, d\theta - (a_1)^3 \int_0^{2\pi} \psi_r (a_1,\theta,t) \, d\theta.$$

Kelvin's circulation theorem implies that

$$\frac{d}{dt}\int_0^{2\pi}\psi_r(a_2,\theta,t)\ d\theta = \frac{d}{dt}\int_0^{2\pi}\psi_r(a_1,\theta,t)\ d\theta = 0,$$

from which it follows that

$$\frac{d}{dt}\sum_{i=1}^2\oint_{\partial\Omega_i} r^2\mathbf{n}_i\cdot\nabla\psi ds = 0. \tag{3.137}$$

This result, together with (3.136), implies that

$$\frac{dM}{dt} = 0 \iff \frac{d}{dt}\iint_\Omega xv - yu\ dxdy = 0.$$

Consequently, we have proved

Theorem 3.8 *The invariance of the Hamiltonian structure for the two dimensional vorticity equation to arbitrary rotations implies that the angular momentum functional*

$$\iint_\Omega xv - yu\ dxdy,$$

is conserved by the dynamics.

3.7 Exercises

Exercise 3.1 *Suppose that $A = A(x,y,t)$ and $B = B(x,y,t)$ are smoothly differentiable functions of their arguments, show that $\partial(A,B)$ can be written in the equivalent forms:*

$$\nabla\cdot[B\mathbf{e}_3\times\nabla(A)]\,, -\nabla\cdot[A\mathbf{e}_3\times\nabla(B)]\,, \ or\ \mathbf{e}_3\cdot(\nabla(A)\times\nabla(B))\,.$$

Exercise 3.2 *Show that the area-integrated kinetic energy*

$$\iint_\Omega \nabla\psi\cdot\nabla\psi\ dxdy,$$

is an invariant of the two dimensional vorticity equation.

Exercise 3.3 *Show that the circulation integrals*

$$\Gamma_i \equiv \oint_{\partial\Omega_i}\mathbf{n}_i\cdot\nabla\psi\ ds,$$

for $i = 1,...,n$ are constants of the motion for the two dimensional vorticity equation, i.e., $\dfrac{\partial\Gamma_i}{\partial t} = 0$. This is Kelvin's circulation theorem.

Exercise 3.4 *Let* $\alpha\left(x,y,t\right)$, $\beta\left(x,y,t\right)$ *and* $\gamma\left(x,y,t\right)$ *be variables defined by the relations*

$$-\frac{p}{\rho} = \gamma_t + \alpha\beta_t + \frac{u^2 + v^2}{2},$$

$$u = \gamma_x + \alpha\left(\beta_x - 1\right),$$

$$v = \gamma_y + \alpha\beta_y,$$

where u, v *and* p *are the velocity components and pressure, respectively, associated with the two-dimensional Euler equations (3.1) through to (3.3). Show that (3.1) through to (3.3) are the first order necessary conditions for an extremal solution to the Lagrangian*

$$\mathcal{L} = \iint_\Omega \gamma_t + \alpha\beta_t + \frac{u^2 + v^2}{2} \ dxdy.$$

Exercise 3.5 *Consider the periodic channel domain given by*

$$\Omega = \left\{(x,y) \mid -y_o < y < y_o, \ 0 < x < L\right\},$$

with the boundary conditions

$$\psi\left(0,y,t\right) = \psi_0,$$

$$\psi\left(L,y,t\right) = \psi_L,$$

and where it is assumed that ψ *is smoothly periodic at* $\pm y_o$. *Determine the corresponding Hamiltonian structure for the two dimensional vorticity equation. Show that the Hamiltonian structure is invariant with respect to arbitrary translations in* y.

Exercise 3.6 *Use Noether's Theorem to show that the invariance of the Hamiltonian structure for the two dimensional vorticity equation with respect to arbitrary translations in* y *in the previous problem implies that the* y-*direction linear momentum functional*

$$\iint_\Omega v \ dxdy,$$

is invariant in time.

Chapter 4

Stability of steady Euler flows

4.1 Steady solutions of the vorticity equation

General steady solutions

Suppose $\varphi_s(x, y)$ is a steady solution of the vorticity equation (3.8). Substitution of $\psi = \varphi_s(x, y)$ into (3.8) leads to

$$\partial\left(\varphi_s, \triangle\varphi_s\right) = 0. \tag{4.1}$$

Besides using the term steady solution, other jargon that is used includes: free modes, equilibrium solutions, fixed points, stationary or time independent solutions. The expression (4.1) states that the Jacobian between the stream function φ_s and the vorticity $\triangle\varphi_s$ with respect to the coordinates x and y is identically zero.

Lemma 4.1 *Suppose $\varphi_s(x, y)$ is a smoothly differentiable steady solution of the vorticity equation (3.8), then*

$$\varphi_s = \Phi\left(\triangle\varphi_s\right), \tag{4.2}$$

where $\Phi\left(*\right)$ is a possibly nonanalytic function of its argument.

   Proof. The lemma is a simple consequence of the properties of Jacobians. We give a simple geometric proof here. Equation (4.1) is equivalent to

$$\partial\left(\varphi_s, \triangle\varphi_s\right) = \frac{\partial\varphi_s}{\partial x}\frac{\partial\triangle\varphi_s}{\partial y} - \frac{\partial\varphi_s}{\partial y}\frac{\partial\triangle\varphi_s}{\partial x}$$

$$= \mathbf{e}_3 \cdot \left[\nabla\varphi_s \times \nabla\left(\triangle\varphi_s\right)\right] = 0.$$

However, this expression implies that the normal vectors to the $\varphi_s - surfaces$ and the $\triangle\varphi_s - surfaces$, given by

$$\frac{\nabla\varphi_s}{|\nabla\varphi_s|} \text{ and } \frac{\nabla\left(\triangle\varphi_s\right)}{|\nabla\left(\triangle\varphi_s\right)|},$$

respectively, are everywhere parallel. This implies that the $\varphi_s - surfaces$ and the $\Delta \varphi_s - surfaces$ coincide, that is,

$$\varphi_s = \Phi \left(\Delta \varphi_s \right). \quad \blacksquare$$

Example 4.2 *Consider the sinusoidal shear flow given by*

$$\varphi_s = \sin(y). \qquad (4.3)$$

It follows that

$$\Delta \varphi_s = \frac{d^2 \varphi_s}{dy^2} = -\sin(y) = -\varphi_s,$$

which implies that

$$\varphi_s = \Phi \left(\Delta \varphi_s \right) \equiv -\Delta \varphi_s.$$

Example 4.3 *Consider the steady gaussian jet given by*

$$\varphi_s = \exp \left(-y^2 \right). \qquad (4.4)$$

It follows that

$$\Delta \varphi_s = \frac{d^2 \varphi_s}{dy^2} = \left(4y^2 - 2 \right) \exp \left(-y^2 \right)$$

$$= -2 \left[2 \ln \left(\varphi_s \right) + 1 \right] \varphi_s \equiv \widetilde{\Phi} \left(\varphi_s \right).$$

which implies that

$$\varphi_s = \Phi \left(\Delta \varphi_s \right),$$

where Φ is the inverse function associated with $\widetilde{\Phi}$.

Vorticity-stream function scatter diagrams

Because arbitrary steady solutions of the vorticity equation (3.8) can be expressed in the relatively simple form (4.2), it is sometimes convenient to depict the steady solution on a *vorticity-stream function scatter diagram*. A vorticity-stream function scatter diagram is simply a two dimensional scatter plot of the pairings

$$\left(\psi \left(x, y, t \right), \ \Delta \psi \left(x, y, t \right) \right),$$

for all $(x, y) \in \Omega$ at a fixed time t (see, for example, Read *et al.* (1986)).

Example 4.4 *The scatter diagram for the sinusoidal flow (4.3) is linear and is given by*

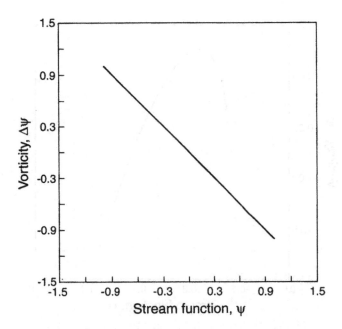

Figure 4.1. Scatter diagram for the sinusoidal shear flow of Example 4.2.

Example 4.5 *The scatter diagram for the gaussian jet (4.4) is nonlinear and is given by*

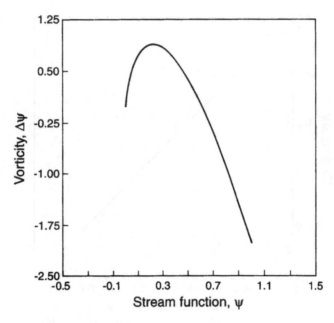

Figure 4.2. Scatter diagram for the sinusoidal shear flow of Example 4.3.

Vorticity-stream function scatter diagrams are a useful diagnostic tool in analyzing experimental data or numerical simulations. Let $\mathcal{A}_{(\psi,\triangle\psi)}$ be the area enclosed by a scatter diagram of

$$(\psi(x,y,t),\ \triangle\psi(x,y,t)).$$

Formally,

$$\mathcal{A}_{(\psi,\triangle\psi)} \equiv \iint_{\mathcal{R}} d\psi d\triangle\psi, \qquad (4.5)$$

where $\mathcal{R}$ is the region of $(\psi, \triangle\psi) - space$ occupied by the scatter diagram. However, we may map the integration region back to $\mathbb{R}^2$ by

$$\mathcal{A}_{(\psi,\triangle\psi)} = \iint_{\tilde{\Omega}} \left|\frac{\partial(\psi,\triangle\psi)}{\partial(x,y)}\right| dxdy \equiv \iint_{\tilde{\Omega}} |\partial(\psi,\triangle\psi)|\ dxdy, \qquad (4.6)$$

where $\widetilde{\Omega}$ is that portion of the spatial domain Ω which is one-to-one with $\mathcal{R}$. Thus, from (4.6) we see immediately that

$$\mathcal{A}_{(\psi,\Delta\psi)} = 0 \Longleftrightarrow |\partial(\psi,\Delta\psi)| = 0. \tag{4.7}$$

Thus we have proved

Theorem 4.6 *Steady solutions of the vorticity equation (3.8) have scatter diagrams with zero area, that is, their scatter diagrams correspond to a possibly multiply-connected curve.*

Example 4.7 *Suppose we slightly modify the gaussian jet (4.4) to*

$$\varphi_s = \exp\left(-y^2\right) + \varepsilon\sin(x),$$

$$\Delta\varphi_s = (4y^2 - 2)\exp\left(-y^2\right) - \varepsilon\sin(x),$$

where $\varepsilon \in \mathbb{R}$. It is straightforward to verify that φ_s is not a steady solution of (4.8) if $\varepsilon \neq 0$ since

$$\partial(\varphi_s, \Delta\varphi_s) = \varepsilon\left(10y - 8y^2\right)\cos(x)\exp\left(-y^2\right) \neq 0.$$

The scatter diagram for $(\varphi_s, \Delta\varphi_s)$ for $\varepsilon = 0.1$ is shown in Fig. 4.3. Observe that the scatter diagram has a nonzero area associated with it. The area of this scatter diagram can be computed using (4.6) as follows

$$\mathcal{A}_{(\varphi_s,\Delta\varphi_s)} = \iint_{\widetilde{\Omega}} |\partial(\varphi_s, \Delta\varphi_s)|\ dxdy$$

$$= |\varepsilon| \int_{-\pi}^{\pi} \int_{-\infty}^{\infty} |(10y - 8y^2)\cos(x)| \exp\left(-y^2\right) dxdy$$

$$= |\varepsilon| \int_{-\pi}^{\pi} |\cos(x)|\, dx \int_{-\infty}^{\infty} |10y - 8y^2| \exp\left(-y^2\right) dy$$

$$= 8|\varepsilon| \left[5 + 4\int_{\frac{5}{4}}^{\infty} \exp(-y^2)dy\right]$$

$$\simeq 84.37 \, |\varepsilon| \neq 0.$$

In the limit $\varepsilon \to 0$ the scatter diagram area goes to zero. Observe that this limit corresponds to φ_s approaching the gaussian jet (4.2) which is a steady solution of the vorticity equation.

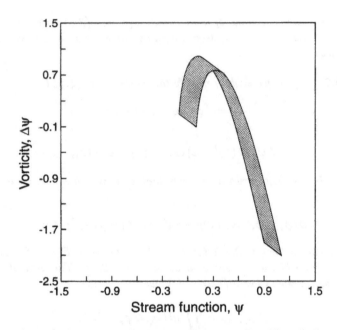

Figure 4.3. Scatter diagram for the nonequilibrium shear flow of Example 4.7.

Theorem 4.6 can be quite useful in analyzing experimental data or numerical simulations of two dimensional flow. For example, if a time series of scatter diagram areas is computed for a given experiment or numerical simulation, then it is possible to qualitatively ascertain the tendency of the observed flow toward or away from a steady solution by examining the tendency of the scatter diagram area to approach or diverge from zero, respectively. This technique has been used, for example, to investigate the possibility that large scale eddy-like disruptions of the mean winds in the northern hemisphere atmosphere, known as *atmospheric blocking events*, correspond to steady solutions of the governing equations (e.g., Butchart *et al.*, 1989; Ek & Swaters, 1994).

Variational principle for steady solutions of the vorticity equation

In contrast to what we saw for the pendulum, steady solutions to the vorticity equation do *not*, in general, satisfy the first order necessary conditions for an extremal of the energy. That is,

$$\frac{\delta H(q_s)}{\delta q} = -\varphi_s \neq 0,$$

where $q_s \equiv \Delta\varphi_s$ (c.f., (3.37) and (3.39); except, of course, for the physically uninteresting trivial solution $\varphi_s = 0$).

However, consider the *constrained* Hamiltonian given by

$$\mathcal{H}(q) \equiv H(q) + C(q), \tag{4.8}$$

where H is the Hamiltonian (3.37) and C is the general Casimir (3.113), respectively. Since H and C are individually invariant, it follows that $\mathcal{H}$ is an invariant of the vorticity equation (3.8). The first variation of $\mathcal{H}$ can be written in the form

$$\delta\mathcal{H}(q) = \iint_\Omega \left(\frac{\delta H}{\delta q} + \frac{\delta C}{\delta q} \right) \delta q \; dxdy$$

$$= \iint_\Omega (C(q) - \psi) \, \delta q \; dxdy. \tag{4.9}$$

It therefore follows that

$$\frac{\delta\mathcal{H}(q)}{\delta q} = C(q) - \psi, \tag{4.10}$$

and in particular observe that

$$\frac{\delta\mathcal{H}(q_s)}{\delta q} = C(q_s) - \varphi_s = 0,$$

if

$$C(q) = \Phi(q),$$

as defined by (4.2). Thus we have proved

Theorem 4.8 *Let $\psi = \varphi_s(x, y)$ be a steady solution of the vorticity equation*

$$\Delta\psi_t + \partial(\psi, \Delta\psi) = 0,$$

satisfying the boundary conditions

$$\varphi_s|_{\partial\Omega_i} = \lambda_i, \quad i = 1, ..., n,$$

with the vorticity-stream function relationship

$$\varphi_s = \Phi\left(\Delta\varphi_s\right),$$

then

$$\delta\mathcal{H}(\Delta\varphi_s) = 0,$$

for

$$\mathcal{H}(q) = \iint\limits_\Omega \left\{ \frac{\nabla\psi\cdot\nabla\psi}{2} + \int\limits^{\Delta\psi} \Phi\left(\xi\right)d\xi \right\} dxdy$$

$$-\sum_{i=1}^n \lambda_i \oint\limits_{\delta\Omega_i} \mathbf{n}\cdot\nabla\psi \, ds, \tag{4.11}$$

where $q = \Delta\psi$.

4.2   Linear stability problem

General linear stability equations

Consider a solution to the vorticity equation in the form

$$\psi\left(x, y, t\right) = \varphi_s(x, y) + \varphi(x, y, t), \tag{4.12}$$

where $\varphi_s(x, y)$ is a steady solution satisfying (4.2). If (4.12) is substituted into (3.8) and all quadratic terms in the perturbation field φ are neglected, it follows that

$$\Delta\varphi_t + \partial\left(\varphi_s, \Delta\varphi\right) + \partial\left(\varphi, \Delta\varphi_s\right) = 0. \tag{4.13}$$

However, since (4.2) holds, this expression can be re-written in the form

$$\Delta\varphi_t = -\partial\left(\Phi\left(\Delta\varphi_s\right), \Delta\varphi\right) - \partial\left(\varphi, \Delta\varphi_s\right)$$

$$= -\Phi'_s\partial\left(\Delta\varphi_s, \Delta\varphi\right) + \partial\left(\Delta\varphi_s, \varphi\right), \tag{4.14}$$

where

$$\Phi'_s \equiv \left[\frac{d\Phi\left(q\right)}{dq}\right]_{q=\Delta\varphi_s}. \tag{4.15}$$

Observing that

$$\partial \left(\triangle \varphi_s, \Phi_s' \triangle \varphi \right) = \Phi_s' \partial \left(\triangle \varphi_s, \triangle \varphi \right) + \triangle \varphi \partial \left(\triangle \varphi_s, \Phi_s' \right)$$

$$= \Phi_s' \partial \left(\triangle \varphi_s, \triangle \varphi \right) + \Phi_s'' \triangle \varphi \partial \left(\triangle \varphi_s, \varphi_s \right)$$

$$= \Phi_s' \partial \left(\triangle \varphi_s, \triangle \varphi \right), \tag{4.16}$$

since $\partial \left(\triangle \varphi_s, \varphi_s \right) = 0$, allows (4.14) to be re-written in the form

$$\triangle \varphi_t + \partial \left(\triangle \varphi_s, \Phi_s' \triangle \varphi - \varphi \right) = 0. \tag{4.17}$$

This is the general linear stability equation associated with the steady solution (4.2).

Perturbation boundary conditions

The linear stability equation (4.17) must be solved with an appropriate initial condition and boundary conditions. Under the assumption that φ_s satisfies the boundary conditions (3.15), then it is appropriate to impose the *natural* perturbation boundary conditions

$$\varphi|_{\partial \Omega_i} = 0, \; i = 1, ..., n. \tag{4.18}$$

If Ω is unbounded we specify the perturbation boundary conditions as they are needed.

4.3 Normal mode equations for parallel shear flows

Our objective in this section is to derive the normal mode equations associated with the linear stability equations for a parallel shear flow and to review Rayleigh's and Fjortoft's stability theorems for these flows. We will show in subsequent sections how the nonlinear Hamiltonian stability theory reduces to these classical linear stability results in the infinitesimally small-amplitude limit. The description given here of the linear theory, although adequate for our purposes, is by no means complete. The interested reader is referred to, for example, Drazin & Reid's (1981) excellent account of the linear stability theory for the Euler equations.

By a parallel shear flow, we mean a solution of the form

$$\psi = \varphi_s(y), \tag{4.19}$$

$$q = q_s(y) = \frac{\partial^2 \varphi_s}{\partial y^2}. \tag{4.20}$$

These are always exact nonlinear steady solutions of (3.8) because

$$\frac{\partial^3 \varphi_s}{\partial t \partial y^2} + \frac{\partial \varphi_s}{\partial x}\frac{\partial^3 \varphi_s}{\partial y^3} - \frac{\partial \varphi_s}{\partial y}\frac{\partial^3 \varphi_s}{\partial x \partial y^2} = 0,$$

since each term is itself identically zero.

The sinusoidal and gaussian flows given by (4.3) and (4.4), respectively, are examples of a parallel shear flow. We have chosen to write these flows as functions of the variable y. This is solely a choice of convenience. Because the vorticity equation is invariant under rotations in $\mathbb{R}^2$ it is always possible to choose the positive x-axis to be oriented in the direction of flow for a parallel shear flow.

The linear stability equation associated with (4.19) can be written in the form

$$\Delta \varphi_t - \frac{\partial^3 \varphi_s}{\partial y^3}\left(\Phi_s' \Delta \varphi - \varphi\right)_x = 0, \tag{4.21}$$

where the vorticity-stream function relationship can be written in the form

$$\varphi_s = \Phi_s \left(\varphi_{s_{yy}}\right). \tag{4.22}$$

The linear stability equation (4.21) can be re-written by noting that it follows from (4.22) that

$$\Phi_s' = \frac{U_s}{U_{s_{yy}}}, \tag{4.23}$$

where the steady x-direction velocity is given by

$$U_s(y) \equiv -\frac{\partial \varphi_s}{\partial y}. \tag{4.24}$$

Substitution of (4.23) and (4.24) into (4.21) leads to

$$\left(\partial_t + U_s \partial_x\right) \Delta \varphi - U_{s_{yy}} \varphi_x = 0, \tag{4.25}$$

which is the usual way of writing the linear stability equation for a parallel shear flow. Equation (4.25) can be expressed in the equivalent form

$$\left(\partial_t + U_s \partial_x\right) \Delta \varphi + \frac{dq_s}{dy} \varphi_x = 0,$$

where $q_s = \varphi_{s_{yy}}(y)$ for the parallel shear flow (4.24). This way of expressing the linear stability equation emphasizes the role the transverse vorticity gradient plays in the second term in (4.25).

For our purposes here we shall only need to consider the periodic channel domain

$$\Omega = \left\{(x, y) \mid -x_o < x < x_o, \ 0 < y < L\right\},$$

with the *perturbation* boundary conditions

$$\varphi(x,0,t) = \varphi(x,L,t) = 0, \tag{4.26}$$

and where it is assumed that $\varphi(x,y,t)$ is smoothly periodic at $x = \pm x_o$ for all $0 \leq y \leq L$ and $t \geq 0$.

The *normal mode* equations are obtained by assuming a solution to (4.25) and (4.26) of the form

$$\varphi(x,y,t) = \varphi_o(y) \exp[ik(x - ct)] + c.c., \tag{4.27}$$

where $k = \dfrac{n\pi}{x_o}$, $n = 0,1,...$, is the real-valued x-direction wavenumber, c is the complex-valued phase speed and $c.c.$ denotes complex-conjugate. Physically, the real-valued solution (4.27) corresponds to a wave travelling in the x-direction which may or may not be exponentially growing in time.

The justification for assuming a solution of the form (4.27) derives from the fact that the solution to the initial boundary-value problem (4.25) and (4.26) with a general initial condition can be solved using a Fourier transform with respect to x and a Laplace transform with respect to t. The travelling wave solutions (4.27) correspond exactly to those contributions to the general solution which arise from the poles in the inverse Laplace-Fourier transform calculation. Thus, those contributions to the solution of the general initial boundary-value problem which exponentially grow in time will correspond to the normal mode solutions for which $Im(kc_I) > 0$ where $c \equiv c_R + ic_I$.

Substitution of (4.27) into (4.25) and (4.26) yields

$$(U_s - c)\left(\varphi_{o_{yy}} - k^2\varphi_o\right) - U_{s_{yy}}\varphi_o = 0, \tag{4.28}$$

$$\varphi_o(0) = \varphi_o(L) = 0. \tag{4.29}$$

This is a homogeneous nonself-adjoint second-order two-point boundary value problem for which nontrivial solutions exist, in general, only for certain discrete relationships between the phase speed and the x-direction wavenumber k. Observe that (4.28) and (4.29) is invariant under the transformation $k \to -k$. Thus we may, without loss of generality, assume that $k \geq 0$. It is a classical result (e.g., Drazin & Reid, 1981) that if the pair $(\varphi_o, c(k))$, for a given wavenumber, is a solution to (4.28) and (4.29), then so is $(\varphi_o^*, c^*(k))$ where the asterisk denotes complex conjugate. Thus, if the phase speed is complex-valued (i.e., has nonzero imaginary part), then there necessarily exist complex pair solutions.

This point is important because it means that the situation will never arise in which all the normal mode perturbations associated with a given parallel shear flow will exponentially decay in time. Consequently, the only type of normal mode stability that can occur is neutral stability in which the time dependent perturbations take the form of nonamplifying waves propagating in the x-direction. In the next two subsections we give two well-known theorems which provide sufficient conditions for neutral stability with respect to normal mode perturbations.

Rayleigh's inflection point theorem

Let us assume that instability occurs so that $c_I \neq 0$. It follows from (4.28) that

$$\varphi_{o_{yy}} - k^2 \varphi_o - \frac{U_{s_{yy}} \varphi_o}{(U_s - c)} = 0, \tag{4.30}$$

which will be nonsingular, i.e., $U_s - c \neq 0$. If (4.30) is multiplied through by $\varphi_o^*(y)$ and the result integrated with respect to y over the interval $(0, L)$, it follows that

$$\int_0^L \left|\varphi_{o_y}\right|^2 + k^2 \left|\varphi_o\right|^2 + \frac{(U_s - c^*) U_{s_{yy}} \left|\varphi_o\right|^2}{\left|U_s - c\right|^2} \, dy = 0, \tag{4.31}$$

where we have integrated the first term by parts exploiting the boundary conditions (4.29).

The imaginary part of (4.31) is given by

$$c_I \left\{ \int_0^L \frac{U_{s_{yy}} \left|\varphi_o\right|^2}{\left|U_s - c\right|^2} \, dy \right\} = 0. \tag{4.32}$$

From (4.32) we immediately see that if $U_{s_{yy}}$ is definite, then all nontrivial normal modes are neutral, i.e., $c_I = 0$. Hence the definiteness of the vorticity gradient $U_{s_{yy}}$ is a *sufficient* condition for stability. Conversely, a *necessary* condition for instability is that there exists at least one $y_* \in (0, L)$ for which $U_{s_{yy}}(y_*) = 0$, i.e., an inflection point must exist in the velocity profile. Thus we have proved

Theorem 4.9 *(Rayleigh, 1880) A necessary condition for normal mode instability is that the parallel shear flow profile $U_s(y)$ contains a point of inflection.*

Fjortoft's stability theorem

Fjortoft's stability theorem exploits the real part of (4.31), given by,

$$\int_0^L \left|\varphi_{o_y}\right|^2 + k^2 \left|\varphi_o\right|^2 + \frac{(U_s - c_R) U_{s_{yy}} \left|\varphi_o\right|^2}{\left|U_s - c\right|^2} \, dy = 0, \tag{4.33}$$

to establish the following result. Suppose instability occurs, then from the Rayleigh inflection point theorem we have

$$c_R \left\{ \int_0^L \frac{U_{s_{yy}} |\varphi_o|^2}{|U_s - c|^2} \, dy \right\} = 0. \tag{4.34}$$

Adding (4.33) and (4.34) together gives

$$\int_0^L \left|\nabla \varphi_{o_y}\right|^2 + k^2 \left|\nabla \varphi_o\right|^2 + \frac{U_s U_{s_{yy}} |\varphi_o|^2}{|U_s - c|^2} \, dy = 0. \tag{4.35}$$

Hence a necessary condition for instability is that there exists $y_* \in (0, L)$ for which

$$U_s U_{s_{yy}} < 0.$$

Conversely, even if the flow contains an inflection point, if

$$U_s U_{s_{yy}} \geq 0,$$

throughout the flow, it follows that (4.35) cannot be satisfied and thus the flow must be neutrally stable. Thus we have proved

Theorem 4.10 *(Fjortoft, 1950) If $U_s U_{s_{yy}} \geq 0$ throughout the flow, then the parallel shear flow $U_s(y)$ is stable to normal mode perturbations.*

It should be pointed out that Fjortoft's theorem as it was originally stated (e.g., see Drazin and Reid, 1981) is that a sufficient condition for stability is that

$$(U_s - U_*) U_{s_{yy}} \geq 0,$$

throughout the flow where U_* is the flow velocity at an inflection point. However, since the vorticity equation is invariant under Galilean transformations we may, without loss of generality, assume a coordinate system for which $U_* = 0$.

4.4 Linear stability theorems

In this section we give Arnol'd's (1965) generalization of Fjortoft's Theorem for arbitrary steady solutions to the vorticity equation. Our approach, in analogy with the approach taken in Section 2.8 for the pendulum, will be to establish conditions for the definiteness of the second variation of the constrained Hamiltonian (4.11) evaluated at the steady solution. Our first task, however, is to show that the second variation is an invariant of the general linear stability equations (4.15).

The second variation of $\mathcal{H}$ can be written in the form

$$\delta^2 \mathcal{H}(q) = \iint\limits_{\Omega} (\Phi(q) - \psi) \delta^2 q + \Phi'(q) (\delta q)^2 + \nabla \delta \psi \cdot \nabla \delta \psi \, dx dy,$$

where $\Phi'(q) \equiv \dfrac{d\Phi}{dq}$. Thus

$$\delta^2 \mathcal{H}(q_s) = \iint\limits_{\Omega} \Phi'_s(q_s) (\delta q)^2 + \nabla \delta \psi \cdot \nabla \delta \psi \, dx dy. \tag{4.36}$$

Theorem 4.11 *The second variation $\delta^2 \mathcal{H}(q_s)$, written in the form,*

$$\delta^2 \mathcal{H}(q_s) = \iint\limits_{\Omega} \Phi'_s(q_s) (\Delta \varphi)^2 + \nabla \varphi \cdot \nabla \varphi \, dx dy,$$

is an invariant of the linear stability equation (4.15).

Proof. We proceed directly. We have

$$\frac{\partial \delta^2 \mathcal{H}(q_s)}{\partial t} = 2 \iint\limits_{\Omega} \Phi'_s(q_s) \Delta \varphi \Delta \varphi_t + \nabla \varphi \cdot \nabla \varphi_t \, dx dy$$

$$= 2 \iint\limits_{\Omega} \Phi'_s(q_s) \Delta \varphi \Delta \varphi_t - \varphi \Delta \varphi_t \, dx dy. \tag{4.37}$$

We will examine this integral term by term. Substitution of (4.15) into the second term in (4.37) implies

$$\iint\limits_{\Omega} \varphi \Delta \varphi_t \, dx dy = -\iint\limits_{\Omega} \varphi \partial (\Delta \varphi_s, \Phi'_s \Delta \varphi - \varphi) \, dx dy$$

$$= -\iint\limits_{\Omega} \varphi \partial (\Delta \varphi_s, \Phi'_s \Delta \varphi) - \frac{1}{2} \partial (\Delta \varphi_s, \varphi^2) \, dx dy$$

$$= -\iint\limits_{\Omega} \varphi \partial (\Delta \varphi_s, \Phi'_s \Delta \varphi) \, dx dy, \tag{4.38}$$

where we have integrated by parts and used the boundary condition (4.16).

For the first term in the right-hand-side of (4.37) we have

$$\iint_\Omega \Phi_s'(q_s)\Delta\varphi\Delta\varphi_t \; dxdy$$

$$= -\iint_\Omega \Phi_s'(q_s)\Delta\varphi\partial\left(\Delta\varphi_s, \Phi_s'\Delta\varphi - \varphi\right) \; dxdy$$

$$= \iint_\Omega \Phi_s'(q_s)\Delta\varphi\partial\left(\Delta\varphi_s, \varphi\right) - \frac{1}{2}\partial\left(\Delta\varphi_s, (\Phi_s'\Delta\varphi)^2\right) \; dxdy. \qquad (4.39)$$

The last integral in (4.39) integrates to zero because

$$\iint_\Omega \partial\left(\Delta\varphi_s, (\Phi_s'\Delta\varphi)^2\right) \; dxdy$$

$$= \iint_\Omega \nabla \cdot \left[(\mathbf{e}_3 \times \nabla\Delta\varphi_s)(\Phi_s'\Delta\varphi)^2\right] \; dxdy$$

$$= -\sum_{i=1}^n \oint_{\partial\Omega_i} \frac{\partial\Delta\varphi_s}{\partial s}(\Phi_s'\Delta\varphi)^2 \; ds$$

$$= -\sum_{i=1}^n \oint_{\partial\Omega_i} \frac{\partial\varphi_s}{\partial s}(\Delta\varphi)^2 \Phi_s' \; ds = 0, \qquad (4.40)$$

where we have used (4.2) and the boundary conditions (3.15).
Substitution of (4.38), (4.39) and (4.40) into (4.37) implies

$$\frac{\partial\delta^2\mathcal{H}(q_s)}{\partial t} = 2\iint_\Omega \{\varphi\partial\left(\Delta\varphi_s, \Phi_s'\Delta\varphi\right)$$

$$+ \Phi_s'(q_s)\Delta\varphi\partial\left(\Delta\varphi_s, \varphi\right)\} \; dxdy$$

$$= 2\iint_\Omega \partial\left(\Delta\varphi_s, \varphi\Phi_s'\Delta\varphi\right) dxdy = 0. \; \blacksquare$$

The integrand of the second variation $\delta^2 \mathcal{H}(q_s)$ is a quadratic form. Thus, if conditions can be found that ensures that the second variation is definite for all perturbations, then we will be able to prove the linear stability in the sense of Liapunov; thereby eliminating *both* algebraic and normal mode instabilities. To distinguish conditions that simply ensure the definiteness of $\delta^2 \mathcal{H}(q_s)$ from the technical requirements of specifying a norm for Liapunov stability, Holm *et al.* (1985) refer to the former as *formal stability*. Clearly, $\delta^2 \mathcal{H}(q_s)$ will be definite if conditions can be established on $\Phi'_s(q_s)$ such that $\delta^2 \mathcal{H}(q_s)$ is positive or negative definite for all perturbations satisfying the appropriate boundary conditions.

Arnol'd's first linear stability theorem

In infinite dimensions care must be taken to specify the norm associated with the stability argument because stability with respect to one norm does not necessarily imply stability with respect to another norm. This is a consequence of the fact that, unlike finite dimensional vector spaces, all norms are not equivalent in infinite dimensions (which results from the loss of compactness which is described below). Although the stability theorem we present here is usually referred to as *Arnol'd's first linear stability theorem*, this result was in fact established by James Clerk Maxwell over a century earlier in a draft manuscript and in correspondence to William Thomson (Maxwell, 1855a,b).

Definition 4.12 *The steady solution* $\varphi_s(x,y)$ *is said to be linearly stable in the sense of Liapunov with respect to the norm* $\|(*)\|$, *if for every* $\varepsilon > 0$ *there exists* $\delta > 0$ *such that* $\|\delta\varphi_o\| < \delta \implies \|\delta\varphi\| < \varepsilon$ *for all* $t \geq 0$, *where* $\delta\varphi_o(x,y) \equiv \delta\varphi(x,y,0)$ *where* $\delta\varphi(x,y,t)$ *solves the linear stability equation (4.15).*

Arnol'd (1965) established the following result for the linear stability in the sense of Liapunov for the steady solution $\varphi_s(x,y)$.

Theorem 4.13 *Let* $\varphi_s(x,y)$ *be a steady solution of the vorticity equation (3.8) as defined by (4.2) satisfying the variational principle in Theorem 4.8. If*

$$0 < \mu_1 \leq \Phi'_s(q_s) \equiv \frac{d\Phi(q_s)}{dq_s} \leq \mu_2 < \infty, \qquad (4.41)$$

for all $(x,y) \in \Omega$, *where* μ_1 *and* μ_2 *are a strictly positive real numbers, then the steady solution* $\varphi_s(x,y)$ *is linearly stable in the sense of Liapunov with respect to the perturbation energy-enstrophy norm*

$$\|\delta\varphi\|^2 = \iint\limits_{\Omega} \nabla\delta\varphi \cdot \nabla\delta\varphi + (\Delta\delta\varphi)^2 \, dxdy. \qquad (4.42)$$

Proof. Clearly, (4.41) is sufficient to ensure that

$$\delta^2 \mathcal{H}(q_s) > 0,$$

for all nontrivial perturbations. All that remains is to establish the following *a priori* estimate. We have, assuming (4.41) holds,

$$\|\delta\varphi\|^2 = \iint_\Omega \nabla\delta\varphi \cdot \nabla\delta\varphi + (\Delta\delta\varphi)^2 \, dxdy$$

$$\leq \Gamma_1^{-1} \iint_\Omega \nabla\delta\varphi \cdot \nabla\delta\varphi + \Phi_s'(q_s) (\Delta\delta\varphi)^2 \, dxdy$$

$$= \Gamma_1^{-1} \delta^2 \mathcal{H}(q_s) = \Gamma_1^{-1} \left[\delta^2 \mathcal{H}(q_s)\right]_{t=0}$$

$$= \Gamma_1^{-1} \iint_\Omega \nabla\delta\varphi_o \cdot \nabla\delta\varphi_o + \Phi_s'(q_s)(\Delta\delta\varphi_o)^2 \, dxdy \leq \frac{\Gamma_2}{\Gamma_1} \|\delta\varphi_o\|^2,$$

where the invariance of $\delta^2\mathcal{H}(q_s)$ has been exploited, and where

$$\Gamma_1 = \min(1, \mu_1) > 0,$$

$$\Gamma_2 = \max(1, \mu_2) > 0,$$

and where $\delta\varphi_o(x,y) \equiv \delta\varphi(x,y,0)$.

Therefore, for every $\varepsilon > 0$, $\|\delta\varphi_o\| < \varepsilon (\Gamma_1/\Gamma_2)^{\frac{1}{2}} \implies \|\delta\varphi\| < \varepsilon$ for all $t \geq 0$. The linear stability of $\varphi_s(x,y)$ in the sense of Liapunov with respect to the energy-enstrophy norm (4.42) has been established provided (4.41) holds. ∎

The stability conditions (4.41) explicitly require that $\inf_\Omega [\Phi_s'(q_s)]$ can be bounded away from zero. If this is not possible, then linear stability in the sense of Liapunov can still be proven with respect to the *energy* norm for $t \geq 0$ provided $0 \leq \Phi_s'(q_s) < \infty$. To establish this result, we first modify the definition of Liapunov stability in the following manner (e.g., Benjamin, 1972).

Definition 4.14 *The steady solution $\varphi_s(x,y)$ is said to be linearly stable in the sense of Liapunov with respect to the norms $\|(*)\|_I$ and $\|(*)\|_{II}$, if for every $\varepsilon > 0$ there exists $\delta > 0$ such that $\|\delta\varphi_o\|_I < \delta \implies \|\delta\varphi\|_{II} < \varepsilon$ for all $t \geq 0$, where $\delta\varphi_o(x,y) \equiv \delta\varphi(x,y,0)$ where $\delta\varphi(x,y,t)$ solves the linear stability equation (4.15).*

Theorem 4.15 *Let $\varphi_s(x,y)$ be a steady solution of the vorticity equation (3.8) as defined by (4.2) satisfying the variational principle in Theorem 4.8. If*

$$0 \leq \Phi'_s(q_s) \equiv \frac{d\Phi(q_s)}{dq_s} \leq \mu < \infty, \qquad (4.43)$$

for all $(x,y) \in \Omega$, where μ is a strictly positive real number, then the steady solution $\varphi_s(x,y)$ is linearly stable in the sense of Liapunov with respect to the perturbation energy-enstrophy norm

$$\|\delta\varphi\|_I^2 = \iint_\Omega \nabla\delta\varphi \cdot \nabla\delta\varphi + (\Delta\delta\varphi)^2 \, dxdy, \qquad (4.44)$$

and the energy norm

$$\|\delta\varphi\|_{II}^2 = E(\delta\varphi) \equiv \iint_\Omega \nabla\delta\varphi \cdot \nabla\delta\varphi \, dxdy. \qquad (4.45)$$

Proof. Clearly, (4.43) is sufficient to ensure that $\delta^2\mathcal{H}(q_s) > 0$ for all nontrivial perturbations. All that remains is to establish the following *a priori* estimate. We have, assuming (4.43) holds,

$$\|\delta\varphi\|_{II}^2 = \iint_\Omega \nabla\delta\varphi \cdot \nabla\delta\varphi \, dxdy$$

$$\leq \delta^2\mathcal{H}(q_s) = \left[\delta^2\mathcal{H}(q_s)\right]_{t=0} \leq \Gamma \|\delta\varphi_o\|_I^2,$$

where $\Gamma = \max(1,\mu) > 0$. Thus, for every $\varepsilon > 0$, $\|\delta\varphi_o\|_I < \varepsilon\Gamma^{-\frac{1}{2}} \implies \|\delta\varphi\|_{II} < \varepsilon$ for all $t \geq 0$. Thus, the linear stability of $\varphi_s(x,y)$ in the sense of Liapunov with respect to the norms (4.44) and (4.45) has been established provided (4.43) holds. ∎

Reduction to Fjortoft's theorem

Theorem 4.15 reduces to Fjortoft's Stability Theorem 4.10 for a parallel shear flow $\varphi_s = \varphi_s(y)$. From (4.2), assuming $\varphi_s = \varphi_s(y)$, it follows that

$$U_s(y) = -\frac{d\varphi_s}{dy} = -\Phi'_s(q_s)\frac{d^3\varphi_s}{dy^3} = \Phi'_s(q_s)U_{s_{yy}}(y),$$

which implies

$$\Phi'_s = \frac{U_s}{U_{s_{yy}}}.$$

Thus, for the parallel shear flow $\varphi_s = \varphi_s(y)$, it follows from (4.43) that

$$\Phi'_s = \frac{U_s}{U_{s_{yy}}} \geq 0 \implies U_s U_{s_{yy}} \geq 0,$$

which is exactly the condition required in Fjortoft's Theorem.

Arnol'd's second linear stability theorem

Arnol'd's first linear stability theorems represent sufficient conditions for the positive definiteness of $\delta^2\mathcal{H}(q_s)$. Arnol'd's second linear stability theorem gives conditions for the negative definiteness of $\delta^2\mathcal{H}(q_s)$. It heuristically follows from (4.36) that if

$$\delta^2\mathcal{H}(q_s) < 0,$$

for all nontrivial perturbations $\delta\varphi$, then $\Phi'_s(q_s)$ must be sufficiently negative so that, roughly speaking, the contribution from the perturbation enstrophy dominates the contribution from the perturbation energy in the integrand of $\delta^2\mathcal{H}(q_s)$. It will, in general, be only possible to establish conditions for the negative definiteness of $\delta^2\mathcal{H}(q_s)$ if the perturbation stream function satisfies a *Poincaré Inequality* (e.g., Ladyzhenskaya, 1969).

Lemma 4.16 *(Poincaré Inequality). Suppose that the minimum eigenvalue, denoted* $\lambda_{\min}$, *of the boundary-value problem*

$$\triangle\phi + \lambda\phi = 0, \ (x,y) \in \Omega, \tag{4.46}$$

$$\phi|_{\partial\Omega} = 0, \tag{4.47}$$

is strictly positive. If $h(x,y)$ *is any twice-continuously differentiable function satisfying* $h|_{\partial\Omega} = 0$, *then*

$$\iint\limits_{\Omega} \nabla h \cdot \nabla h \ dxdy \le \frac{1}{\lambda_{\min}} \iint\limits_{\Omega} (\triangle h)^2 \ dxdy. \tag{4.48}$$

Proof. There are many ways of establishing this result. Perhaps the easiest is to exploit a *Rayleigh-Ritz* variational principle for the lowest eigenvalue (e.g., Zauderer, 1983). If (4.46) is multiplied through by $\triangle\phi$ and the second term integrated by parts, it follows that

$$\lambda = \frac{\iint\limits_{\Omega} (\triangle\phi)^2 \ dxdy}{\iint\limits_{\Omega} \nabla\phi \cdot \nabla\phi \ dxdy}.$$

Thus a *Rayleigh-Ritz* variational principle for $\lambda_{\min}$ is

$$\lambda_{\min} = \min_{\phi} \left\{ \frac{\iint\limits_{\Omega} (\triangle\phi)^2 \ dxdy}{\iint\limits_{\Omega} \nabla\phi \cdot \nabla\phi \ dxdy} \right\}, \tag{4.49}$$

where the minimization is done over all twice-continuously differentiable nontrivial functions $\phi(x, y)$ which satisfy the boundary condition (4.47). It follows from (4.49) that if $h(x, y)$ is an arbitrary twice-continuously differentiable function satisfying $h|_{\partial\Omega} = 0$, then

$$\frac{\displaystyle\iint_\Omega (\triangle h)^2 \, dxdy}{\displaystyle\iint_\Omega \nabla h \cdot \nabla h \, dxdy} \geq \lambda_{\min},$$

which can be re-arranged into (4.48). ■

The Poincaré inequality (4.48) is only useful if the minimum eigenvalue can be bounded away from zero. A sufficient condition for $\lambda_{\min} > 0$ is that the domain Ω be bounded in at least one direction in the plane. For example, if $\Omega = \mathbb{R}^2$, which is not bounded in any direction, the bounded eigenfunctions of Laplace's operator are proportional to $\exp\left[ikx + ily\right]$ implying that $\lambda = k^2 + l^2$ which cannot be bounded away from zero.

However, for the periodic channel domain

$$\Omega = \left\{(x, y) \mid -x_o < x < x_o, \ 0 < y < L\right\},$$

the nontrivial eigenfunctions are proportional to

$$\exp\left[\pm\frac{in\pi x}{x_o}\right] \sin\left[\frac{m\pi y}{L}\right],$$

for $n = 0, 1, ...,$ and $m = 1, 2, ... $, implying that

$$\lambda = \left(\frac{n\pi}{x_o}\right)^2 + \left(\frac{m\pi}{L}\right)^2,$$

which in turn implies

$$\lambda_{\min} \equiv \min_{n,m}\{\lambda\} = \left(\frac{\pi}{L}\right)^2 > 0.$$

Theorem 4.17 *Let $\varphi_s(x, y)$ be a steady solution of the vorticity equation (3.8) as defined by (4.2) satisfying the variational principle in Theorem 4.8. If*

$$-\infty < \mu_1 \leq \Phi_s'(q_s) \equiv \frac{d\Phi(q_s)}{dq_s} \leq \mu_2 < \frac{-1}{\lambda_{\min}} < 0, \qquad (4.50)$$

for all $(x, y) \in \Omega$, where μ_1 and μ_2 are strictly negative real numbers where $\lambda_{\min}$ is the strictly positive minimum eigenvalue of the boundary-value problem

$$\triangle\phi + \lambda\phi = 0, \ (x, y) \in \Omega,$$

$$\phi|_{\partial\Omega} = 0,$$

then the steady solution $\varphi_s(x, y)$ is linearly stable in the sense of Liapunov with respect to the perturbation enstrophy norm

$$\|\delta\varphi\|^2 = \iint_\Omega (\triangle\delta\varphi)^2 \, dxdy. \tag{4.51}$$

Proof. First, we shall bound the perturbation norm (4.51) in terms of $\delta^2\mathcal{H}(q_s)$. We have

$$\delta^2\mathcal{H}(q_s) = \iint_\Omega \nabla\delta\varphi \cdot \nabla\delta\varphi + \Phi'_s(q_s)(\triangle\delta\varphi)^2 \, dxdy$$

$$\leq \iint_\Omega \left(\frac{1}{\lambda_{\min}} + \Phi'_s(q_s)\right)(\triangle\delta\varphi)^2 \, dxdy \leq \Gamma_1 \|\delta\varphi\|^2 < 0, \tag{4.52}$$

where $\Gamma_1 \equiv \lambda_{\min}^{-1} + \mu_2 < 0$. This inequality may be re-arranged into the form

$$\|\delta\varphi\|^2 \leq \Gamma_1^{-1}\delta^2\mathcal{H}(q_s) = \Gamma_1^{-1}\left[\delta^2\mathcal{H}(q_s)\right]_{t=0}$$

$$= \Gamma_1^{-1} \iint_\Omega \nabla\delta\varphi_o \cdot \nabla\delta\varphi_o + \Phi'_s(q_s)(\triangle\delta\varphi_o)^2 \, dxdy$$

$$\leq \Gamma_1^{-1} \iint_\Omega \Phi'_s(q_s)(\triangle\delta\varphi_o)^2 \, dxdy \leq \Gamma_2 \|\delta\varphi_o\|^2,$$

where $\Gamma_2 \equiv \mu_1\Gamma_1^{-1} > 0$. Therefore, for every $\varepsilon > 0$, $\|\delta\varphi_o\| < \varepsilon\Gamma_2^{-\frac{1}{2}} \implies \|\delta\varphi\| < \varepsilon$ for all $t \geq 0$. The linear stability of $\varphi_s(x, y)$ in the sense of Liapunov with respect to the enstrophy norm (4.51) has been established provided (4.50) holds. ∎

Summary of the linear stability argument for a steady solution

The underlying mathematical procedure for a proof of *linear* stability in the sense of Liapunov based on the Hamiltonian structure of a given dynamical system can be summarized with the following four steps:

1. For given system of partial differential equations determine an underlying Hamiltonian formulation. Care must be taken that the proposed Poisson bracket satisfies all the required algebraic properties. If the Hamiltonian formulation is noninvertible, i.e., the kernel of the Poisson bracket contains nontrivial functionals, one must usually determine the complete family of Casimir functionals.

2. Determine a variational principle for the steady solutions of the system of governing equations. If the Hamiltonian formulation is canonical it may be possible to show that the steady solutions satisfy the first order necessary conditions for an extremal of the Hamiltonian. If the Hamiltonian formulation is noninvertible it will usually be the case that the variational principle will need to be based on an appropriately constrained Hamiltonian where the constraints will be written in terms of the Casimirs. As it turns out, the above procedure is usually sufficient for obtaining a variational principle for a *steady* solution of a system of equations. However it may be necessary to impose additional constraints on a the Hamiltonian to obtain a variational principle for other types of solutions. For example, in Chapters 5 and 6 it will be shown that *steadily-travelling* solutions can be described by a variational principle associated with a Hamiltonian constrained by (Casimirs if needed and) the linear momentum invariant.

3. Determine conditions for the definiteness of the second variation of the functional used in the variational principle evaluated at the steady solution. The second variation of this functional evaluated at the steady solution will always be an invariant of the appropriate linear stability equations. Thus if the second variation of the variational functional evaluated at the steady solution is definite, then the steady solution will be *formally* stable.

4. The second variation of the variational functional evaluated at the steady solution will always be a homogeneous functional of degree two with respect to the perturbation field. This property, together with the formal stability conditions, can be exploited to bound the second variation of the variational functional evaluated at the steady solution in terms of norms on the perturbation field. This bound and the invariance of the second variation of the variational functional evaluated at the steady solution can be combined to establish the linear stability in the sense of Liapunov of the steady solution with respect to the bounding norms.

4.5   Nonlinear stability theorems

The loss of compactness in infinite dimensions

The outline of the fluid dynamical stability argument we have used up until this point is essentially identical to the stability argument developed for the pendulum in Chapter 2. However, the nonlinear fluid dynamical stability argument must be modified to take into account topological problems associated with an infinite dimensional phase space. The geometric idea behind the nonlinear stability argument remains the same in that we wish to show that the equilibrium solution corresponds to a strict extremum of the (constrained) Hamiltonian.

In finite dimensions, the definiteness of the second differential of the Hamiltonian (2.41) (for which the variational principle Theorem 2.2 holds), is sufficient to show that the steady solution corresponds to a strict extremum of the Hamiltonian. For example, it is an exercise in single variable calculus to show that if $f(x)$ is a twice-continuously differentiable function on $\mathbb{R}$ satisfying $f'(a) = 0$ and $f''(a) > 0$, then there exists $\varepsilon > 0$ such that $f(x) > f(a)$ for all x in the interval $0 < |x - a| < \varepsilon$.

In infinite dimensions, the definiteness of $\delta^2 \mathcal{H}(q_s)$ is not, in general, sufficient to guarantee the existence of $\varepsilon > 0$ such that $\mathcal{H}(q)$ can be bounded away from $\mathcal{H}(q_s)$ for all $q(x, y, t)$ satisfying $0 < \|q - q_s\| < \varepsilon$. This is a consequence of the fact that in infinite dimensions not every convergent sequence on the unit ball converges to a point on the unit ball, i.e., infinite dimensional vector spaces are not compact.

This point can be easily illustrated with the following simple example. Let ℓ denote the space of absolutely summable infinite sequences and let $\beta = (\beta_1, ..., \beta_n, ...)$ be an element of ℓ. Let $\Upsilon(\beta)$ be the functional defined by

$$\Upsilon(\beta) = \frac{1}{2} \sum_{i=1}^{\infty} \beta_i^2. \tag{4.53}$$

Note that $\Upsilon(\beta)$ is finite for every $\beta \in \ell$ since

$$\Upsilon(\beta) = \frac{1}{2} \sum_{i=1}^{\infty} \beta_i^2 = \frac{1}{2} \sum_{i=1}^{\infty} \left\{ |\beta_i| \sum_{j=1}^{\infty} \delta_{ij} |\beta_j| \right\}$$

$$\leq \frac{1}{2} \sum_{i=1}^{\infty} \left\{ |\beta_i| \sum_{j=1}^{\infty} |\beta_j| \right\} = \frac{1}{2} \left[\sum_{i=1}^{\infty} |\beta_i| \right]^2 < \infty.$$

The first and second variations of $\Upsilon(\beta)$ are given by, respectively,

$$\delta\Upsilon(\beta) = \sum_{i=1}^{\infty} \beta_i \delta\beta_i,$$

$$\delta^2\Upsilon(\beta) = \sum_{i=1}^{\infty} \beta_i \delta^2\beta_i + \sum_{i=1}^{\infty} (\delta\beta_i)^2.$$

Thus we see that

$$\Upsilon(0) = 0, \tag{4.54}$$

$$\delta\Upsilon(0) = 0, \tag{4.55}$$

$$\delta^2 \Upsilon(0) = \sum_{i=1}^{\infty} (\delta\beta_i)^2 > 0. \tag{4.56}$$

We now show with a counter-example that (4.54) through to (4.56) are not sufficient to ensure that $\Upsilon(\beta) > \Upsilon(0)$ for $0 < \|\beta\|_\ell < \varepsilon$ no matter how small $\varepsilon > 0$ is. Consider the sequence of ℓ-sequences given by $\{X_n\}_{n=1}^{\infty}$ where

$$X_n \equiv \mu \left(\overbrace{n^{-1}, ..., n^{-1}}^{n \text{ copies}}, 0, 0, ... \right),$$

and $\mu = \dfrac{\varepsilon}{2} > 0$. Note that

$$\|X_n\|_\ell = \mu \sum_{i=1}^{n} \left| \frac{1}{n} \right| = \mu > 0,$$

$$\Upsilon(X_n) = \sum_{i=1}^{n} \frac{\mu^2}{n^2} = \frac{\mu^2}{n}.$$

However, since

$$\lim_{n \to \infty} \Upsilon(X_n) = 0 = \Upsilon(0),$$

even though $\lim_{n \to \infty} \|X_n\|_\ell = \mu = \dfrac{\varepsilon}{2} < \varepsilon$, it follows that $\Upsilon(\beta)$ *cannot* be bounded away from $\Upsilon(0)$ for every β satisfying $0 < \|\beta\|_\ell < \varepsilon$. This example serves to make the point that additional hypotheses must be introduced on the constrained Hamiltonian to ensure that $\mathcal{H}(q)$ can be bounded away from $\mathcal{H}(q_s)$ for all $q(x, y, t)$ satisfying $0 < \|q - q_s\| < \varepsilon$ for some $\varepsilon > 0$.

Arnol'd's first nonlinear stability theorem

As in the linear stability results it is important to specify the stability norm in the stability argument.

Definition 4.18 *The steady solution $\varphi_s(x, y)$ is said to be nonlinearly stable in the sense of Liapunov with respect to the norm $\|(*)\|$, if for every $\varepsilon > 0$ there exists $\delta > 0$ such that $\|\psi_o - \varphi_s\| < \delta \implies \|\psi - \varphi_s\| < \varepsilon$ for all $t \geq 0$, where $\psi_o(x, y) \equiv \psi(x, y, 0)$ where $\psi(x, y, t)$ solves the nonlinear vorticity equation (3.8).*

Let us introduce the functional $\mathcal{N}(q)$ given by

$$\mathcal{N}(q) = \mathcal{H}(q_s + q) - \mathcal{H}(q_s) - \delta\mathcal{H}(q_s), \tag{4.57}$$

where $\mathcal{H}(q)$ is the functional associated with the variational principle in Theorem 4.8 and where $\delta\mathcal{H}(q_s)$ is written in the form

$$\delta\mathcal{H}(q_s) = \iint\limits_{\Omega} (\Phi(q_s) - \psi_s)\, q\, dxdy = 0. \tag{4.58}$$

The *total* vorticity, given by $q_T(x,y,t) \equiv q_s(x,y) + q(x,y,t)$, solves the full nonlinear vorticity equation (3.8). We will call $q(x,y,t)$ the disturbance or perturbation vorticity.

The functional $\mathcal{N}(q)$ is an invariant of the vorticity equation (3.8) since $\mathcal{H}(q_T)$ is an invariant, $\mathcal{H}(q_s)$ is time independent and $\delta\mathcal{H}(q_s) = 0$. Consequently,

$$\mathcal{N}(q) = \mathcal{N}(\tilde{q}), \tag{4.59}$$

where $\tilde{q}(x,y) \equiv q(x,y,0)$. In addition, observe that $\mathcal{N}(0) = 0$. The nonlinear stability argument will center on establishing conditions on $\mathcal{H}$ so that $\mathcal{N}(q)$ can be bounded away from $\mathcal{N}(0)$ for all $q(x,y,t)$ satisfying $0 < \|q\| < \varepsilon$ for some $\varepsilon > 0$.

If $\mathcal{N}(q)$ is Taylor expanded about $q = 0$, it follows that

$$\mathcal{N}(q) = \frac{1}{2}\delta^2\mathcal{H}(q_s) + \; h.o.t., \tag{4.60}$$

where the variations in $\delta^2\mathcal{H}(q_s)$ are replaced with the disturbance vorticity $q(x,y,t)$. Thus, to leading order, $\mathcal{N}(q)$ is proportional to the second variation of the constrained Hamiltonian $\mathcal{H}$. Observe that if $\mathcal{H}$ is at most quadratic in the vorticity (i.e., Φ is linear with respect to its argument or, equivalently, C is proportional to the enstrophy) , then $\delta^n\mathcal{H}(q_s) = 0$ for $n \geq 3$ and thus $\mathcal{N}(q) = \frac{1}{2}\delta^2\mathcal{H}(q_s)$. In this special circumstance the $\delta^2\mathcal{H}(q_s)$ is an invariant of the full nonlinear vorticity equation and the definiteness of $\delta^2\mathcal{H}(q_s)$ does imply that $\mathcal{N}(q)$ can be bounded away from $\mathcal{N}(0)$ for all $q(x,y,t)$ satisfying $0 < \|q\| < \varepsilon$ for some $\varepsilon > 0$.

If we substitute the constrained Hamiltonian $\mathcal{H}$ given by (4.11) into (4.57), $\mathcal{N}(q)$ can be written in the form

$$\mathcal{N}(q) = \iint\limits_{\Omega} \left\{ \frac{\nabla\varphi \cdot \nabla\varphi}{2} \right.$$

$$\left. + \int\limits_{\Delta\varphi_s}^{\Delta\varphi_s+\Delta\varphi} \Phi(\xi)\, d\xi - \Phi(\Delta\varphi_s)\,\Delta\varphi \right\} dxdy, \tag{4.61}$$

where $q_s \equiv \Delta\varphi_s$ and $q \equiv \Delta\varphi$. To establish conditions ensuring that $\mathcal{N}(q)$ can be bounded away from $\mathcal{N}(0)$, we need the following result.

Lemma 4.19 *Suppose* $F : C^{(2)}(\mathbb{R}) \to \mathbb{R}$. *If there exist finite real numbers* α *and* β *for which*

$$\alpha \le F''(\xi) \le \beta, \tag{4.62}$$

for all $\xi \in \mathbb{R}$, *then*

$$\frac{\alpha h^2}{2} \le F(x+h) - F(x) - F'(x) h \le \frac{\beta h^2}{2}, \tag{4.63}$$

for all x *and* h *in* $\mathbb{R}$.

Proof. The inequality (4.62) can be integrated once with respect to ξ over the interval $\xi \in (y, x)$ to give

$$\alpha(y - x) \le F'(y) - F'(x) \le \beta(y - x).$$

And this inequality can be integrated with respect to y in the form

$$\alpha \int_x^{x+h} (y - x)\, dy \le \int_x^{x+h} [F'(y) - F'(x)]\ dy$$

$$\le \beta \int_x^{x+h} (y - x)\, dy,$$

which, if evaluated, gives (4.63). ∎

As an application of this Lemma, suppose that $F(q)$ is given by

$$F(q) = \int^q \Phi(\xi)\, d\xi,$$

$$F'(q) = \Phi(q),$$

$$F''(q) = \Phi'(q).$$

Then, if there exist real numbers α and β such that

$$\alpha \le \Phi'(q) \le \beta,$$

for all q, it follows from Lemma 4.19 that

$$\frac{\alpha(\Delta\varphi)^2}{2} \le \int_{\Delta\varphi_s}^{\Delta\varphi_s + \Delta\varphi} \Phi(\xi)\, d\xi - \Phi(\Delta\varphi_s)\, \Delta\varphi \le \frac{\beta(\Delta\varphi)^2}{2}, \tag{4.64}$$

for all $\Delta\varphi_s$ and $\Delta\varphi$.

Theorem 4.20 *Let $\varphi_s(x, y)$ be a steady solution of the vorticity equation (3.8) as defined by $\varphi_s = \Phi(\triangle\varphi_s)$ satisfying the variational principle in Theorem 4.8. If*

$$0 < \mu_1 \le \Phi'(q) \le \mu_2 < \infty, \tag{4.65}$$

for all q, where μ_1 and μ_2 are a strictly positive real numbers, then the steady solution $\varphi_s(x, y)$ is nonlinearly stable in the sense of Liapunov with respect to the disturbance energy-enstrophy norm

$$\|\varphi\|^2 = \iint\limits_{\Omega} \nabla\varphi \cdot \nabla\varphi + (\triangle\varphi)^2 \, dxdy. \tag{4.66}$$

Proof. Lemma 4.19 together with (4.65) implies that

$$\frac{\mu_1(\triangle\varphi)^2}{2} \le \int\limits_{\triangle\varphi_s}^{\triangle\varphi_s+\triangle\varphi} \Phi(\xi) \, d\xi - \Phi(\triangle\varphi_s)\triangle\varphi \le \frac{\mu_2(\triangle\varphi)^2}{2},$$

for all $\triangle\varphi$. It therefore follows that $\mathcal{N}(q)$ satisfies

$$0 < \iint\limits_{\Omega} \nabla\varphi \cdot \nabla\varphi + \mu_1(\triangle\varphi)^2 \, dxdy \le 2\mathcal{N}(q)$$

$$= 2\mathcal{N}(\tilde{q}) \le \iint\limits_{\Omega} \nabla\tilde{\varphi} \cdot \nabla\tilde{\varphi} + \mu_2(\triangle\tilde{\varphi})^2 \, dxdy,$$

where $\tilde{\varphi}(x, y) = \varphi(x, y, 0)$. From this inequality we conclude that

$$\|\varphi\| \le \Gamma \|\tilde{\varphi}\|,$$

where

$$\Gamma = \left[\frac{\max(1, \mu_2)}{\min(1, \mu_1)}\right]^{\frac{1}{2}} > 0.$$

Thus, for every $\varepsilon > 0$, $\|\tilde{\varphi}\| < \varepsilon\Gamma^{-1} \implies \|\varphi\| < \varepsilon$ for all $t \ge 0$. The nonlinear stability of $\varphi_s(x, y)$ has been established with respect to the energy-enstrophy disturbance norm assuming (4.65) holds. ∎

Similar to what we saw in the linear stability Theorem 4.13, the stability conditions (4.65) explicitly require that μ_1 is strictly greater than zero. If this is not possible, then nonlinear stability in the sense of Liapunov can still be proven provided we modify the definition of nonlinear Liapunov stability in a manner similar to Definition 4.14.

Definition 4.21 *The steady solution $\varphi_s(x, y)$ is said to be nonlinearly stable in the sense of Liapunov with respect to the norms $\|(*)\|_I$ and $\|(*)\|_{II}$, if for every $\varepsilon > 0$ there exists $\delta > 0$ such that $\|\psi_o - \varphi_s\|_I < \delta \Longrightarrow \|\psi - \varphi_s\|_{II} < \varepsilon$ for all $t \geq 0$, where $\psi_o(x, y) \equiv \psi(x, y, 0)$ where $\psi(x, y, t)$ solves the nonlinear vorticity equation (3.8).*

Theorem 4.22 *Let $\varphi_s(x, y)$ be a steady solution of the vorticity equation (3.8) as defined by $\varphi_s = \Phi(\Delta\varphi_s)$ satisfying the variational principle in Theorem 4.8. If*

$$0 \leq \Phi'(q) \leq \mu < \infty, \tag{4.67}$$

for all q, where μ is a strictly positive real number, then the steady solution $\varphi_s(x, y)$ is nonlinearly stable in the sense of Liapunov with respect to the disturbance energy-enstrophy norm

$$\|\varphi\|_I^2 = \iint\limits_{\Omega} \nabla\varphi \cdot \nabla\varphi + (\Delta\varphi)^2 \, dxdy, \tag{4.68}$$

and the disturbance energy norm

$$\|\varphi\|_{II}^2 = E(\varphi) \equiv \iint\limits_{\Omega} \nabla\varphi \cdot \nabla\varphi \, dxdy. \tag{4.69}$$

Proof. The hypothesis (4.67) and Lemma 4.19 imply

$$0 \leq \int\limits_{\Delta\varphi_s}^{\Delta\varphi_s + \Delta\varphi} \Phi(\xi)\, d\xi - \Phi(\Delta\varphi_s)\, \Delta\varphi \leq \frac{\mu(\Delta\varphi)^2}{2},$$

for all $\Delta\varphi$. It therefore follows that $\mathcal{N}(q)$ satisfies

$$0 < \iint\limits_{\Omega} \nabla\varphi \cdot \nabla\varphi \, dxdy \leq 2\mathcal{N}(q)$$

$$= 2\mathcal{N}(\tilde{q}) \leq \iint\limits_{\Omega} \nabla\tilde{\varphi} \cdot \nabla\tilde{\varphi} + \mu(\Delta\tilde{\varphi})^2 \, dxdy,$$

where $\tilde{\varphi}(x, y) = \varphi(x, y, 0)$. From this inequality we conclude that

$$\|\varphi\|_{II} \leq \Gamma \|\tilde{\varphi}\|_I,$$

where

$$\Gamma = [\max(1, \mu)]^{\frac{1}{2}} > 0.$$

Thus, for every $\varepsilon > 0$, $\|\tilde{\varphi}\|_I < \varepsilon\Gamma^{-1} \Longrightarrow \|\varphi\|_{II} < \varepsilon$ for all $t \geq 0$ and the nonlinear stability of $\varphi_s(x, y)$ has been established provided (4.67) holds. ∎

Arnol'd's second nonlinear stability theorem

We complete this section by giving the nonlinear generalization of Theorem 4.17. Before we do this, however, it is necessary to comment on the disturbance boundary conditions. If we assume that the total stream function (composed of the sum of the steady *and* perturbation stream function) satisfy the boundary conditions (3.15), then it is appropriate to impose the *natural* finite-amplitude disturbance boundary conditions

$$\varphi = 0 \ on \ \partial\Omega. \tag{4.70}$$

In what follows these are the boundary conditions assumed.

Theorem 4.23 *Let $\varphi_s(x,y)$ be a steady solution of the vorticity equation (3.8) as defined by $\varphi_s = \Phi(\Delta\varphi_s)$ satisfying the variational principle in Theorem 4.8. If*

$$-\infty < \mu_1 \le \Phi'(q) \le \mu_2 < \frac{-1}{\lambda_{\min}} < 0, \tag{4.71}$$

for all q, where μ_1 and μ_2 are strictly negative real numbers where $\lambda_{\min}$ is the strictly positive minimum eigenvalue of the boundary-value problem

$$\Delta\phi + \lambda\phi = 0, \ (x,y) \in \Omega,$$

$$\phi|_{\partial\Omega} = 0,$$

then the steady solution $\varphi_s(x,y)$ is linearly stable in the sense of Liapunov with respect to the disturbance enstrophy norm

$$\|\varphi\|^2 = \iint\limits_\Omega (\Delta\varphi)^2 \, dxdy. \tag{4.72}$$

Proof. First, we shall bound the perturbation norm (4.72) in terms of $\mathcal{N}(q)$. We have

$$\mathcal{N}(q) = \iint\limits_\Omega \frac{\nabla\varphi \cdot \nabla\varphi}{2} + \int\limits_{\Delta\varphi_s}^{\Delta\varphi_s + \Delta\varphi} \Phi(\xi) \, d\xi - \Phi(\Delta\varphi_s) \, \Delta\varphi \ dxdy$$

$$\le \frac{1}{2} \iint\limits_\Omega \left(\frac{1}{\lambda_{\min}} + \mu_2\right)(\Delta\varphi)^2 \ dxdy \le \frac{\Gamma_1}{2} \|\varphi\|^2 < 0, \tag{4.73}$$

where $\Gamma_1 \equiv \lambda_{\min}^{-1} + \mu_2 < 0$ and where we have explicitly used Lemma 4.19 and the natural boundary conditions (4.70). The inequality (4.73) may be re-arranged into the form

$$\|\delta\varphi\|^2 \leq 2\Gamma_1^{-1}\mathcal{N}(q) = 2\Gamma_1^{-1}\mathcal{N}(\tilde{q})$$

$$= \Gamma_1^{-1} \iint_\Omega \nabla\tilde{\varphi} \cdot \nabla\tilde{\varphi} + 2 \int_{\Delta\varphi_s}^{\Delta\varphi_s + \Delta\tilde{\varphi}} \Phi(\xi) \, d\xi - 2\Phi(\Delta\varphi_s)\, \Delta\varphi \; dxdy$$

$$\leq \Gamma_1^{-1} \iint_\Omega \nabla\tilde{\varphi} \cdot \nabla\tilde{\varphi} + \mu_1 (\Delta\tilde{\varphi})^2 \; dxdy$$

$$\leq \Gamma_2 \iint_\Omega (\Delta\tilde{\varphi})^2 \, dxdy \leq \Gamma_2 \|\tilde{\varphi}\|^2 \,,$$

where $\Gamma_2 \equiv \mu_1\Gamma_1^{-1} > 0$ and where $\tilde{\varphi}(x,y) = \varphi(x,y,0)$. Therefore, for every $\varepsilon > 0$, $\|\tilde{\varphi}\| < \varepsilon\Gamma_2^{-\frac{1}{2}} \implies \|\varphi\| < \varepsilon$ for all $t \geq 0$. The nonlinear stability of $\varphi_s(x,y)$ has been established provided (4.71) holds. ∎

Summary of the nonlinear stability argument for a steady solution

The underlying mathematical procedure for a proof of *nonlinear* stability in the sense of Liapunov based on the Hamiltonian structure of a given dynamical system can be summarized with the following four steps:

1. Complete steps one through to four of the linear stability argument given at the end of Section 4.4.

2. Introduce the nonlinear functional $\mathcal{N}$ given by (4.57) where $\mathcal{H}$ is the functional associated with the variational principle describing the steady solutions.

3. Determine the appropriate convexity hypotheses that the density associated with the functional $\mathcal{N}$ must satisfy to ensure that $\mathcal{N}$ is definite. These estimates will usually lead to the introduction of a norm on the perturbation fields required to prove nonlinear stability in the sense of Liapunov.

4. Exploit the definiteness of $\mathcal{N}$ and the convexity hypotheses to establish the required estimates for nonlinear stability in the sense of Liapunov.

4.6   Andrews's Theorem

Up until now we have not addressed the issue of the existence of flows that can satisfy
the stability theorems of Section 4.4 and 4.5. These theorems would be of little
physical utility if the only flows which can satisfy the conditions of the theorems
are trivial ones. As we shall see, *Andrews's Theorem* (Andrews, 1984; Carnevale
& Shepherd, 1990; Chern & Marsden, 1990) highlights that there is a relationship
between underlying symmetries of the Hamiltonian formulation and the class of flows
that can, in principle, satisfy the Arnol'd stability theorems.

It is, however, very important to make clear that Andrews's Theorem does
not imply that those flows which cannot satisfy the Arnol'd stability conditions are
unstable. The stability conditions are sufficient and not necessary. Flows which do
not satisfy the conditions for Arnol'd stability may, in fact, be stable. It would be
our contention, however, that in this situation further analysis is required in developing appropriate bounds on the second variation of the, for example, constrained
Hamiltonian. We will have more to say on this point in our treatment of the stability
of the KdV soliton in Chapter 6.

To make our discussion concrete in this section, we will work in the periodic
channel

$$\Omega = \{(x,y) \,|\, -x_o < x < x_o, \, 0 < y < L\}, \tag{4.74}$$

with the boundary conditions

$$\psi(x,0,t) = \psi_0, \tag{4.75}$$

$$\psi(x,L,t) = \psi_L, \tag{4.76}$$

and periodicity conditions

$$\left[\frac{\partial^{n+m}\psi}{\partial x^n \partial y^m}\right]_{x=-x_o} = \left[\frac{\partial^{n+m}\psi}{\partial x^n \partial y^m}\right]_{x=x_o}, \tag{4.77}$$

for all $n \geq 0$ and $m \geq 0$.

If follows from Theorem 4.8 that the steady solution $\psi = \varphi_s(x,y)$ of the
vorticity equation (3.8) with the vorticity-stream function relationship

$$\varphi_s = \Phi(\Delta\varphi_s),$$

satisfies

$$\delta\mathcal{H}(\Delta\varphi_s) = 0,$$

for

$$\mathcal{H}(q) = \iint_{\Omega} \left\{ \frac{\nabla\psi \cdot \nabla\psi}{2} + \int^{\Delta\psi} \Phi(\xi)\, d\xi \right\} dxdy$$

$$+\psi_0 \int_{-x_o}^{x_o} \psi_y(x,0,t)\, dx - \psi_L \int_{-x_o}^{x_o} \psi_y(x,L,t)\, dx, \tag{4.78}$$

where $q \equiv \Delta\psi$.

Theorems 4.13 and 4.14 establish that $\varphi_s(x,y)$ is linearly stable in the sense of Liapunov if there exists a positive real number μ for which

$$0 \le \Phi'_s < \mu < \infty, \ \forall\ (x,y) \in \Omega. \tag{4.79}$$

Theorem 4.15 establishes that $\varphi_s(x,y)$ is linearly stable in the sense of Liapunov if there exists negative real numbers μ_1 and μ_2 for which

$$-\infty < \mu_1 \le \Phi'_s \le \mu_2 < -(L/\pi)^2 < 0, \ \forall\,(x,y) \in \Omega. \tag{4.80}$$

We now give a characterization of the class of flows that can, in principle, satisfy these stability conditions.

Theorem 4.24 *(Andrews's Theorem) The only steady solutions for the periodic channel domain Ω given by (4.74) satisfying the boundary and periodicity conditions (4.75) through to (4.77) which can satisfy the linear stability criteria (4.79) or (4.80) are the parallel shear flows $\varphi_s = \varphi_s(y)$.*

    Proof. Let $\varphi_s(x,y)$ be an arbitrary steady solution to the vorticity equation (3.8) given by $\varphi_s = \Phi(\Delta\varphi_s)$, it follows that

$$\frac{\partial\varphi_s}{\partial x} = \Phi'_s \frac{\partial\Delta\varphi_s}{\partial x}.$$

If this expression is multiplied through by $\dfrac{\partial\Delta\varphi_s}{\partial x}$ and the result integrated over Ω we obtain

$$\iint_{\Omega} \nabla\left(\frac{\partial\varphi_s}{\partial x}\right) \cdot \nabla\left(\frac{\partial\varphi_s}{\partial x}\right) + \Phi'_s \left(\frac{\partial\Delta\varphi_s}{\partial x}\right)^2 dxdy$$

$$= \int_{-x_o}^{x_o} \left[\frac{\partial\varphi_s}{\partial x}\frac{\partial^2\varphi_s}{\partial x\partial y}\right]_{y=L} - \left[\frac{\partial\varphi_s}{\partial x}\frac{\partial^2\varphi_s}{\partial x\partial y}\right]_{y=0} dx$$

$$+ \int_0^L \left[\frac{\partial \varphi_s}{\partial x} \frac{\partial^2 \varphi_s}{\partial x^2} \right]_{x=x_o} - \left[\frac{\partial \varphi_s}{\partial x} \frac{\partial^2 \varphi_s}{\partial x^2} \right]_{x=-x_o} dy, \qquad (4.81)$$

where we have integrated by parts. However, exploiting the periodicity of φ_s and

$$\left[\frac{\partial \varphi_s}{\partial x} \right]_{y=0} = \left[\frac{\partial \varphi_s}{\partial x} \right]_{y=L} = 0,$$

it follows from (4.81) that

$$\iint_\Omega \nabla \left(\frac{\partial \varphi_s}{\partial x} \right) \cdot \nabla \left(\frac{\partial \varphi_s}{\partial x} \right) + \Phi'_s \left(\frac{\partial \Delta \varphi_s}{\partial x} \right)^2 \, dx dy = 0. \qquad (4.82)$$

However, if

$$0 \leq \Phi'_s < \mu < \infty, \ \forall \ (x,y) \in \Omega,$$

it follows from (4.82) that

$$\nabla \left(\frac{\partial \varphi_s}{\partial x} \right) = 0 \Longrightarrow \frac{\partial \varphi_s}{\partial x} = 0 \Longrightarrow \varphi_s = \varphi_s(y),$$

where (4.75) and (4.76) have been used. Thus we have shown that if (4.79) holds then the only flows in the periodic channel defined by (4.74) that can satisfy the conditions of Theorem 4.13 are parallel shear flows of the form $\varphi_s = \varphi_s(y)$.

Assuming that φ_s is sufficiently smooth, it follows that $\dfrac{\partial \varphi_s}{\partial x}$ satisfies the Poincaré Inequality (Lemma 4.15 applied to the periodic channel)

$$\iint_\Omega \nabla \left(\frac{\partial \varphi_s}{\partial x} \right) \cdot \nabla \left(\frac{\partial \varphi_s}{\partial x} \right) \, dx dy \leq \left(\frac{L}{\pi} \right)^2 \iint_\Omega \left(\frac{\partial \Delta \varphi_s}{\partial x} \right)^2 \, dx dy.$$

This inequality together with (4.82) implies

$$0 \leq \iint_\Omega \left[\Phi'_s + \left(\frac{L}{\pi} \right)^2 \right] \left(\frac{\partial \Delta \varphi_s}{\partial x} \right)^2 \, dx dy. \qquad (4.83)$$

But if

$$-\infty < \mu_1 \leq \Phi'_s \leq \mu_2 < (L/\pi)^2 < 0, \ \forall \ (x,y) \in \Omega$$

(i.e., the conditions of Theorem 4.41 hold), then

$$-\infty < \Phi'_s + \left(\frac{L}{\pi} \right)^2 < 0, \forall \ (x,y) \in \Omega.$$

This inequality and (4.83) implies that

$$0 \leq \iint_{\Omega} \left[\Phi'_s + \left(\frac{L}{\pi} \right)^2 \right] \left(\frac{\partial \Delta \varphi_s}{\partial x} \right)^2 \, dx dy \leq 0.$$

This can only hold if

$$\frac{\partial \Delta \varphi_s}{\partial x} = 0,$$

$\forall \, (x, y) \in \Omega$. If this expression is multiplied through by $\dfrac{\partial \varphi_s}{\partial x}$ and the result integrated over Ω we obtain

$$\iint_{\Omega} \nabla \left(\frac{\partial \varphi_s}{\partial x} \right) \cdot \nabla \left(\frac{\partial \varphi_s}{\partial x} \right) \, dx dy = 0,$$

where we have exploited the periodicity of φ_s at $x = \pm x_o$ and the boundary conditions on $y = 0$ and $y = L$. It follows from this expression that $\varphi_s = \varphi_s(y)$. Thus we have proven that the only flows which can satisfy the conditions of Theorems 4.13 and 4.14 in the periodic channel domain (4.74) are the parallel shear flows $\varphi_s = \varphi_s(y)$. ∎

Symmetry restrictions , such as that just described, are not exclusive to the periodic channel domain. Suppose we consider smooth solutions to the vorticity equation (3.8) on the unbounded domain $\Omega = \mathbb{R}^2$ such that the area-integrated energy and enstrophy given by, respectively,

$$\iint_{\mathbb{R}^2} \nabla \psi \cdot \nabla \psi \, dx dy < \infty, \tag{4.84}$$

$$\iint_{\mathbb{R}^2} (\Delta \psi)^2 \, dx dy < \infty, \tag{4.85}$$

are finite. An example of the type of solutions that would satisfy (4.84) and (4.85) are isolated eddy solutions with compact support.

Theorem 4.25 *The only steady (smooth) solutions for the unbounded domain $\mathbb{R}^2$ with the property $\varphi_s \to 0$ as $x^2 + y^2 \to \infty$ that can satisfy the linear stability condition (4.79) are the trivial solutions.*

Proof. Let $\varphi_s(x, y)$ be an arbitrary steady solution (with finite energy and enstrophy) to the vorticity equation (3.8) given by $\varphi_s = \Phi(\Delta \varphi_s)$, it follows that

$$\nabla \varphi_s = \Phi'_s \nabla (\Delta \varphi_s).$$

If the inner product between this expression and $\nabla\left(\triangle\varphi_s\right)$ is formed, we obtain

$$\iint\limits_{\Omega} \nabla\left(\frac{\partial\varphi_s}{\partial x}\right)\cdot\nabla\left(\frac{\partial\varphi_s}{\partial x}\right) + \nabla\left(\frac{\partial\varphi_s}{\partial y}\right)\cdot\nabla\left(\frac{\partial\varphi_s}{\partial y}\right)$$

$$+\Phi'_s\left[\nabla\left(\triangle\varphi_s\right)\cdot\nabla\left(\triangle\varphi_s\right)\right]\,dxdy = 0,$$

where we have integrated by parts and used the boundary conditions.

However, if

$$0\le\Phi'_s<\mu<\infty,\ \forall\ (x,y)\in\Omega,$$

it follows that

$$\nabla\left(\varphi_s\right)=\mathbf{0}\implies\varphi_s=\text{constant}=0,$$

where $\varphi_s\to 0$ as $x^2+y^2\to\infty$ has been used. ∎

4.7 Flows with special symmetries

Flows with special symmetries may have alternate variational principles associated with them which can be exploited to establish stability. The two examples considered here are parallel shear flows of the form $\varphi_s=\varphi_s\left(y\right)$ and circular flows of the form $\varphi_s=\varphi_s\left(r\right)$ where $r=\left(x^2+y^2\right)^{\frac{1}{2}}$.

Alternate variational principle for parallel shear flows

In Section 4.4 we showed how Arnol'd's first linear stability theorem (Theorem 4.15) reduced to Fjortoft's stability theorem (Theorem 4.10) for a parallel shear flow. In this section we show that Rayleigh's inflection point theorem (Theorem 4.9) implies that the second variation of the area-integrated x-direction linear momentum constrained by an appropriate Casimir functional is definite for the parallel shear flow $\varphi_s=\varphi_s\left(y\right)$.

To make our discussion concrete in this subsection, we will work in the periodic channel

$$\Omega = \{(x,y)\,|-x_o<x<x_o,\ 0<y<L\},\tag{4.86}$$

with the boundary conditions

$$\psi\left(x,0,t\right)=\psi_0,\tag{4.87}$$

$$\psi\left(x,L,t\right)=\psi_L,\tag{4.88}$$

and where it is assumed that $\psi(x, y, t)$ is smoothly periodic at $x = \pm x_0$ where $\psi(x, y, t)$ solves the vorticity equation

$$\Delta\psi_t + \partial(\psi, \Delta\psi) = 0. \tag{4.89}$$

It is easy to see that $\psi = \varphi_s(y)$ is an exact nonlinear solution provided $\varphi_s(y)$ satisfies (4.87) and (4.88).

The area-integrated x-direction linear momentum is given by (modulo a weighted conserved circulation integral on $\partial\Omega$; see Section 3.6)

$$M(q) = \iint_\Omega yq \, dxdy = \iint_\Omega y\Delta\psi \, dxdy. \tag{4.90}$$

We have already seen that the invariance in time of $M(q)$ follows from Noether's Theorem applied to the invariance of the Hamiltonian structure to arbitrary translations with respect to x. One can also show that $M(q)$ is an invariant directly, i.e.,

$$\frac{\partial M}{\partial t} = \iint_\Omega y\Delta\psi_t \, dxdy$$

$$= -\iint_\Omega y\partial(\psi, \Delta\psi) \, dxdy = -\iint_\Omega y\nabla\cdot(\mathbf{u}\Delta\psi) \, dxdy$$

$$= -\oint_{\partial\Omega} y(\mathbf{n}\cdot\mathbf{u})\Delta\psi \, ds + \iint_\Omega \psi_x\Delta\psi \, dxdy$$

$$= \iint_\Omega \nabla\cdot(\psi_x\nabla\psi) - \frac{1}{2}(\nabla\psi\cdot\nabla\psi)_x \, dxdy = 0,$$

where $\mathbf{u} = (-\psi_y, \psi_x)$, and where we have repeatedly used the boundary and periodicity conditions.

Consider the invariant functional $\mathcal{L}(q)$ given by a linear combination of the x-direction momentum and a Casimir in the form

$$\mathcal{L}(q) = \iint_\Omega \left\{ \int^q C(\xi) \, d\xi - yq \right\} dxdy$$

$$= \iint_\Omega \left\{ \int^{\Delta\psi} C(\xi) \, d\xi - y\Delta\psi \right\} dxdy. \tag{4.91}$$

This functional is clearly invariant since it is a linear combination of two invariant functionals.

The first variation of $\mathcal{L}$ can be written in the form

$$\delta\mathcal{L}(q) = \iint\limits_{\Omega} \{\mathcal{C}(q) - y\}\, \delta q\, dxdy.$$

We want to determine the form of $\mathcal{C} = \mathcal{C}(q)$ so that $\delta\mathcal{L}(\varphi_s(y)) = 0$. Let $\mathcal{Y}(q_s)$ be the inverse function associated with $q_s = \varphi_{s_{yy}}(y)$, i.e.,

$$\mathcal{Y}(q_s(y)) = y.$$

Thus we have the variational principle

Theorem 4.26 *Let $\psi = \varphi_s(y)$ be a parallel shear flow solution of the vorticity equation (4.89) in the periodic channel domain (4.86) satisfying the boundary conditions (4.87) and (4.88) which is smoothly periodic at $x = \pm x_o$, then*

$$\delta\mathcal{L}(q_s) = 0, \tag{4.92}$$

for

$$\mathcal{L}(q) = \iint\limits_{\Omega}\left\{\int^{q}\mathcal{Y}(\xi)\,d\xi - yq\right\}\,dxdy, \tag{4.93}$$

where $\mathcal{Y}(q)$ is the inverse function associated with $q = \varphi_{s_{yy}}(y)$.

Conditions which can establish the linear stability in the sense of Liapunov are obtained from $\delta^2\mathcal{L}(q_s)$ given by

$$\delta^2\mathcal{L}(q_s) = \iint\limits_{\Omega}\mathcal{Y}'(q_s)\,(\Delta\delta\psi)^2\;dxdy, \tag{4.94}$$

where

$$\mathcal{Y}'(q_s) \equiv \left[\frac{d\mathcal{Y}(q)}{dq}\right]_{q=q_s=\varphi_{s_{yy}}}$$

Clearly, $\delta^2\mathcal{L}(q_s)$ will be definite if $\mathcal{Y}'(q_s)$ is definite.

Lemma 4.27 *$\delta^2\mathcal{L}(q_s)$ is an invariant of the linear stability equation associated with $\varphi_s(y)$.*

Proof. The argument is similar to that used to prove Theorem 4.11. We have

$$\frac{\partial \delta^2 \mathcal{L}(q_s)}{\partial t} = 2 \iint\limits_{\Omega} \mathcal{Y}'(q_s)\, \Delta \delta \psi_t\, \Delta \delta \psi\, dxdy$$

$$= -2 \iint\limits_{\Omega} \mathcal{Y}'(q_s)\, \Delta \delta \psi\, \{\partial(\varphi_s, \Delta \delta \psi) + \partial(\delta \psi, q_s)\}\, dxdy$$

$$= \iint\limits_{\Omega} \left\{ \left[\mathcal{Y}'(q_s)\, \varphi_{s_y}\, (\Delta \delta \psi)^2\right]_x - 2 \mathcal{Y}'(q_s)\, q_{s_y} \delta \psi_x \Delta \delta \psi \right\} dxdy$$

$$= \iint\limits_{\Omega} \left[(\nabla \psi \cdot \nabla \psi)_x - 2\nabla \cdot (\delta \psi_x \nabla \delta \psi) \right] dxdy = 0$$

where the periodicity and boundary conditions have been used and where $\mathcal{Y}'(q_s)\, q_{s_y} = 1$ has been used. ∎

Rayleigh's Inflection Point Theorem for parallel shear flows via $\delta^2 \mathcal{L}(q_s)$

Theorem 4.28 *The parallel shear solution $\psi = \varphi_s(y)$ (in the periodic channel domain (4.86) satisfying the boundary conditions (4.87) and (4.88) and the smooth periodicity conditions at $x = \pm x_o$) is linearly stable in the sense of Liapunov with respect to the perturbation enstrophy norm*

$$\|\delta \psi\|^2 = \iint\limits_{\Omega} (\Delta \delta \psi)^2\, dxdy,$$

if $\mathcal{Y}'(q_s)$ is of definite sign.

Proof. Let us first suppose that $\mathcal{Y}'(q_s)$ satisfies

$$0 < \Gamma_1 \leq \mathcal{Y}'(q_s) \leq \Gamma_2 < \infty, \tag{4.95}$$

where Γ_1 and Γ_2 are strictly-positive real numbers. It follows that

$$\|\delta \psi\|^2 = \iint\limits_{\Omega} (\Delta \delta \psi)^2\, dxdy$$

$$\leq \frac{1}{\Gamma_1} \iint\limits_{\Omega} \mathcal{Y}'(q_s)\, (\Delta \delta \psi)^2\, dxdy = \frac{1}{\Gamma_1} \delta^2 \mathcal{L}(q_s)$$

$$= \frac{1}{\Gamma_1} \left[\delta^2 \mathcal{L}\left(q_s\right)\right]_{t=0} = \frac{1}{\Gamma_1} \iint\limits_{\Omega} \mathcal{Y}'\left(q_s\right) \left(\Delta\delta\widetilde{\psi}\right)^2 \, dx dy$$

$$\leq \frac{\Gamma_2}{\Gamma_1} \iint\limits_{\Omega} \left(\Delta\delta\widetilde{\psi}\right)^2 \, dx dy = \frac{\Gamma_2}{\Gamma_1} \left\|\delta\widetilde{\psi}\right\|^2,$$

where $\delta\widetilde{\psi}(x,y) = \delta\psi(x,y,0)$. Thus for every $\varepsilon > 0 \; \exists \; \delta = \varepsilon \left(\Gamma_1/\Gamma_2\right)^{\frac{1}{2}} > 0$ such that $\left\|\delta\widetilde{\psi}\right\| \leq \delta \Longrightarrow \|\delta\psi\| \leq \varepsilon \; \forall \; t \geq 0$ if (4.95) holds.

On the other hand if $\mathcal{Y}'\left(q_s\right)$ satisfies

$$-\infty < \Gamma_1 \leq \mathcal{Y}'\left(q_s\right) \leq \Gamma_2 < 0, \tag{4.96}$$

where Γ_1 and Γ_2 are strictly negative real numbers, it follows that

$$\|\delta\psi\|^2 = \iint\limits_{\Omega} \left(\Delta\delta\psi\right)^2 \, dx dy$$

$$\leq \frac{1}{\Gamma_2} \iint\limits_{\Omega} \mathcal{Y}'\left(q_s\right) \left(\Delta\delta\psi\right)^2 \, dx dy = \frac{1}{\Gamma_2} \delta^2 \mathcal{L}\left(q_s\right)$$

$$= \frac{1}{\Gamma_2} \left[\delta^2 \mathcal{L}\left(q_s\right)\right]_{t=0} = \frac{1}{\Gamma_2} \iint\limits_{\Omega} \mathcal{Y}'\left(q_s\right) \left(\Delta\delta\widetilde{\psi}\right)^2 \, dx dy$$

$$\leq \frac{\Gamma_1}{\Gamma_2} \iint\limits_{\Omega} \left(\Delta\delta\widetilde{\psi}\right)^2 \, dx dy = \frac{\Gamma_1}{\Gamma_2} \left\|\delta\widetilde{\psi}\right\|^2.$$

Thus for every $\varepsilon > 0 \; \exists \; \delta = \varepsilon \left(\Gamma_2/\Gamma_1\right)^{\frac{1}{2}} > 0$ such that $\left\|\delta\widetilde{\psi}\right\| \leq \delta \Longrightarrow \|\delta\psi\| \leq \varepsilon \; \forall \; t \geq 0$ if (4.96) holds. ∎

Theorem 4.28 is equivalent to Rayleigh's inflection point theorem. The definition of $\mathcal{Y}\left(q_s\right)$ is

$$\mathcal{Y}\left(\varphi_{s_{yy}}(y)\right) = y,$$

from which it follows

$$\mathcal{Y}'\left(q_s\right) \varphi_{s_{yyy}} = 1 \Longleftrightarrow \mathcal{Y}'\left(q_s\right) = -\frac{1}{U_{s_{yy}}}, \quad \left(U_{s_{yy}} \neq 0\right),$$

where $U_s = -\varphi_{s_y}$. It therefore follows that the definiteness of $\mathcal{Y}'\left(q_s\right)$ is equivalent to the definiteness of $U_{s_{yy}}$. Thus, Theorem 4.28 implies that a sufficient condition for the linear stability of the parallel shear flow $\varphi_s(y)$ is that $U_{s_{yy}}$ be definite or, alternatively, that the mean flow profile does *not* contain an inflection point in it. This is, of course, just Rayleigh's inflection point theorem.

Nonlinear generalization of Rayleigh's Theorem for parallel shear flows

The nonlinear generalization of Rayleigh's inflection point theorem for a parallel shear flow is given as follows. First, let us introduce the functional $\mathcal{N}(q)$ given by

$$\mathcal{N}(q) = \mathcal{L}(q_s + q) - \mathcal{L}(q_s) - \delta\mathcal{L}(q_s), \qquad (4.97)$$

where $\mathcal{L}(q)$ is given by (4.93) and where $\delta\mathcal{L}(q_s)$ is written in the form

$$\delta\mathcal{L}(q_s) = \iint_\Omega \{\mathcal{Y}(\Delta\varphi_s) - y\}\, q\, dxdy = 0. \qquad (4.98)$$

The *total* stream function, given by

$$\psi(x,y,t) = \varphi_s(y) + \varphi(x,y,t),$$

solves the vorticity equation (4.89). The disturbance stream function is given by $\varphi(x,y,t)$ which satisfies the boundary conditions $\varphi(x,0,t) = \varphi(x,L,t) = 0$ and is assumed to be smoothly periodic at $x = \pm x_o$. Since $\mathcal{N}(q)$ is the sum of three individually invariant functionals, it follows $\mathcal{N}(\varphi)$ is itself an invariant of the nonlinear vorticity equation. Hence

$$\mathcal{N}(q(x,y,t)) = \mathcal{N}(\widetilde{q}(x,y)),$$

where $\widetilde{q}(x,y) = q(x,y,0)$ and $\widetilde{\varphi}(x,y) = \varphi(x,y,0)$.

Substitution of (4.93) into (4.97) yields

$$\mathcal{N}(q) = \iint_\Omega \left\{ \int_{\Delta\varphi}^{\Delta\varphi_s + \Delta\varphi} \mathcal{Y}(\xi)\, d\xi - \mathcal{Y}(\Delta\varphi_s)\, \Delta\varphi \right\} dxdy. \qquad (4.99)$$

If $\mathcal{N}(q)$ is Taylor expanded about $q = 0$, one finds

$$\mathcal{N}(q) = \frac{1}{2}\delta^2\mathcal{L}(q_s) + h.o.t.,$$

where it is understood that the variations δq in (4.94) are replaced with the disturbance vorticity $q(x,y,t)$.

The nonlinear generalization of Rayleigh's inflection point theorem requires assuming that there exist finite real numbers α and β such that

$$\alpha \leq \frac{d\mathcal{Y}(q)}{dq} \leq \beta, \qquad (4.100)$$

for all $q \in \mathbb{R}$. If (4.100) holds, then it follows from Lemma 4.19 that

$$\frac{\alpha(\Delta\varphi)^2}{2} \leq \int_{\Delta\varphi}^{\Delta\varphi_s + \Delta\varphi} \mathcal{Y}(\xi)\, d\xi - \mathcal{Y}(\Delta\varphi_s)\, \Delta\varphi \leq \frac{\beta(\Delta\varphi)^2}{2}, \qquad (4.101)$$

for all disturbance vorticity $\Delta\varphi$.

Theorem 4.29 *If*

$$0 < \alpha \leq \frac{d\mathcal{Y}(q)}{dq} \leq \beta < \infty, \tag{4.102}$$

or if

$$-\infty < \alpha \leq \frac{d\mathcal{Y}(q)}{dq} \leq \beta < 0, \tag{4.103}$$

for all $q \in \mathbb{R}$, then the parallel shear flow solution $\psi = \varphi_s(y)$ is nonlinearly stable in the sense of Liapunov with respect to the disturbance enstrophy norm

$$\|\varphi\|^2 = \iint_{\Omega} (\Delta\varphi)^2 \, dxdy. \tag{4.104}$$

Proof. The argument is straightforward. It follows from (4.99), (4.101), (4.102), (4.103) and the invariance of $\mathcal{N}(q)$ that

$$0 < \alpha \|\varphi\|^2 \leq 2\mathcal{N}(q)$$

$$= 2\mathcal{N}(\tilde{q}) \leq \beta \|\tilde{\varphi}\|^2 < \infty.$$

Thus for every $\varepsilon > 0 \; \exists \; \delta = \varepsilon (\alpha/\beta)^{\frac{1}{2}} > 0$ such that $\|\tilde{\varphi}\| \leq \delta \implies \|\varphi\| \leq \varepsilon \; \forall \; t \geq 0$ if (4.102) holds.

On the other hand if (4.103) holds, then

$$-\infty < \alpha \|\tilde{\varphi}\|^2 \leq 2\mathcal{N}(\tilde{q})$$

$$= 2\mathcal{N}(q) \leq \beta \|\varphi\|^2 < 0,$$

from which it follows that

$$\|\varphi\| \leq \left(\frac{\alpha}{\beta}\right)^{\frac{1}{2}} \|\tilde{\varphi}\|.$$

Thus for every $\varepsilon > 0 \; \exists \; \delta = \varepsilon (\beta/\alpha)^{\frac{1}{2}} > 0$ such that $\|\tilde{\varphi}\| \leq \delta \implies \|\varphi\| \leq \varepsilon \; \forall \; t \geq 0$ if (4.103) holds. ∎

Alternate variational principle for circular flows

For our discussion in this subsection we will restrict attention to the circular domain

$$\Omega = \left\{ (x,y) \mid 0 \le x^2 + y^2 < a \right\}, \tag{4.105}$$

with the boundary condition

$$\psi = 0, \quad on \quad r = a, \tag{4.106}$$

where we have introduced the polar coordinates

$$r = \sqrt{x^2 + y^2}, \quad \tan(\theta) = \frac{y}{x}. \tag{4.107}$$

It will be assumed that the radius a is finite. This assumption can be relaxed provided certain conditions on the stream function at infinity hold. These will be pointed out as they arise.

The alternate variational principle presented here for circular flows of the form $\varphi_s = \varphi_s(r)$ corresponds to showing that these flows satisfy the first-order necessary conditions for an extremal of the angular momentum functional constrained by a Casimir. Our presentation is similar to the description given by Vladimirov (1986) or Carnevale & Shepherd (1990).

The angular momentum (with unit density) associated with the vorticity equation (4.89) may be written in the form (modulo a conserved circulation integral; see Section 3.6)

$$A = \iint_\Omega r^2 \triangle \psi \, rdrd\theta. \tag{4.108}$$

We have already seen that the invariance in time of A follows from Noether's Theorem applied to the invariance of the Hamiltonian structure to arbitrary rotations. However, the invariance in time of A may be directly verified with the calculation

$$\frac{dA}{dt} = \iint_\Omega r^3 \triangle \psi_t \, drd\theta$$

$$= - \iint_\Omega r^3 \nabla \cdot \left[\left(\frac{-\psi_\theta}{r}, \psi_r \right) \triangle \psi \right] drd\theta$$

$$= \iint_\Omega r^2 \left[\frac{\partial}{\partial r} \left(\psi_\theta \triangle \psi \right) - \frac{\partial}{\partial \theta} \left(\psi_r \triangle \psi \right) \right] drd\theta$$

$$= a^2 \int_0^{2\pi} \psi_\theta \triangle \psi d\theta - 2 \iint_\Omega r \psi_\theta \triangle \psi dr d\theta$$

$$= \iint_\Omega r \left[\frac{\partial}{\partial \theta} \left(\nabla \psi \cdot \nabla \psi \right) - 2 \nabla \cdot \left(\psi_\theta \nabla \psi \right) \right] dr d\theta$$

$$= -2 \iint_\Omega \left[\frac{\partial}{\partial r} \left(r \psi_\theta \psi_r \right) + \frac{\partial}{\partial \theta} \left(\frac{\psi_r \psi_\theta}{r} \right) \right] dr d\theta = 0,$$

where we have repeatedly used the boundary condition (4.106) and the required periodicity with respect to θ. If the domain is unbounded, the angular momentum integral will exist (i.e., be finite) provided the vorticity $\triangle \psi$ vanishes sufficiently rapidly at infinity.

An alternate variational principle can be constructed for circular flows via the invariant functional

$$\mathcal{L}(q) = \iint_\Omega \left\{ \int^{\triangle \psi} \mathcal{C}(\xi) \, d\xi - r^2 \triangle \psi \right\} r dr d\theta, \tag{4.109}$$

which can be interpreted as (proportional to) the angular momentum constrained by a Casimir. The first variation is given by

$$\delta \mathcal{L}(q) = \iint_\Omega \left\{ \mathcal{C}(\triangle \psi) - r^2 \right\} \delta q \, r dr d\theta. \tag{4.110}$$

Similar to our work for parallel shear flows we want to determine the Casimir density $\mathcal{C}(\triangle \psi)$ so that $\delta \mathcal{L}(q_s(r)) = 0$. Let $\mathcal{R}(q_s)$ be the inverse function associated with

$$q_s = \triangle \varphi_s(r) = \varphi_{s_{rr}}(r) + \varphi_{s_r}(r)/r,$$

i.e.,

$$\mathcal{R}\left(\varphi_{s_{rr}}(r) + \varphi_{s_r}(r)/r \right) = r.$$

We therefore have the following variational principle

Theorem 4.30 *The circular flow* $\varphi_s(r)$ *satisfies*

$$\delta \mathcal{L}(q_s(r)) = 0, \tag{4.111}$$

for the functional

$$\mathcal{L}(q) = \iint_\Omega \left\{ \int^q \mathcal{R}^2(\xi) \, d\xi - r^2 q \right\} r dr d\theta, \tag{4.112}$$

where $\mathcal{R}(q)$ *is the inverse function associated with* $q = \triangle \varphi_s(r)$.

Conditions for the linear stability of $\varphi_s(r)$ can be obtained from examining the second variation $\delta^2 \mathcal{L}\left(q_s\left(r\right)\right)$, given by

$$\delta^2 \mathcal{L}\left(q_s\left(r\right)\right) = 2 \iint\limits_{\Omega} \mathcal{R}\left(q_s\right) \mathcal{R}'\left(q_s\right) \left(\delta q\right)^2 r dr d\theta$$

$$= 2 \iint\limits_{\Omega} \mathcal{R}'\left(q_s\right) \left(\delta q\right)^2 r^2 dr d\theta, \tag{4.113}$$

where

$$\mathcal{R}'\left(q_s\right) \equiv \left[\frac{d\mathcal{R}\left(q\right)}{dq}\right]_{q=q_s},$$

and we have used the fact that $r = \mathcal{R}\left(q_s\right)$. Clearly, $\delta^2 \mathcal{L}\left(q_s\right)$ will be definite if $\mathcal{R}'\left(q_s\right)$ is.

Lemma 4.31 $\delta^2 \mathcal{L}\left(q_s\right)$ *is an invariant of the linear stability equation associated with* $\varphi_s(r)$.

Proof. The argument is similar to that used to prove Lemma 4.27. It follows from (4.113) that

$$\frac{\partial \delta^2 \mathcal{L}\left(q_s\right)}{\partial t} = 4 \iint\limits_{\Omega} \mathcal{R}'\left(q_s\right) \Delta \delta \psi_t \, \Delta \delta \psi \, r^2 dr d\theta$$

$$= -4 \iint\limits_{\Omega} \mathcal{R}'\left(q_s\right) \Delta \delta \psi \left\{\partial\left(\varphi_s, \Delta \delta \psi\right) + \partial\left(\delta \psi, q_s\right)\right\} r^2 dr d\theta$$

$$= \iint\limits_{\Omega} \left\{4 \mathcal{R}'\left(q_s\right) q_{s_r} \delta \psi_\theta \Delta \delta \psi - 2 \left[\mathcal{R}'\left(q_s\right) \varphi_{s_r} \left(\Delta \delta \psi\right)^2\right]_\theta\right\} r dr d\theta = 0,$$

$$= \iint\limits_{\Omega} \left[4 \nabla \cdot \left(\delta \psi_\theta \nabla \delta \psi\right) - 2 \left(\nabla \psi \cdot \nabla \psi\right)_\theta\right] r dr d\theta = 0,$$

where the periodicity with respect to θ and the boundary condition $\delta \psi\left(a, \theta, t\right) = 0$ have been used, and where

$$\mathcal{R}'\left(q_s\right) q_{s_r} = 1,$$

has been used. ∎

Rayleigh's Theorem for circular flows

We begin our discussion in this subsection by establishing the analogue of Rayleigh's Inflection Point Theorem 4.9 for a circular flow in the circular domain Ω given by (4.105) with boundary condition (4.106). For a circular flow $\varphi_s = \varphi_s(r)$, the linear stability equation (4.13) can be written in the form

$$\Delta\varphi_t + r^{-1}\left(\varphi_{s_r}\Delta\varphi_\theta - \varphi_\theta q_{s_r}\right), \tag{4.114}$$

where we have introduced the polar coordinates (4.107) with

$$q_s \equiv \Delta\varphi_s = r^{-1}\left(r\varphi_{s_r}\right)_r,$$

and

$$\Delta\varphi = r^{-1}\frac{\partial}{\partial r}\left(r\frac{\partial\varphi}{\partial r}\right) + r^{-2}\frac{\partial^2\varphi}{\partial\theta^2},$$

with $\varphi(r,\theta,t)$ the perturbation stream function.. It is assumed that $\varphi|_{\partial\Omega} = 0$. Substitution of the normal mode decomposition

$$\varphi(r,\theta,t) = \varphi_o(r)\exp\left[in\left(\theta - ct\right)\right] + c.c., \tag{4.115}$$

where $n = 0,1,2,\dots$ into the linear stability equation (4.114) leads to the normal mode problem

$$\left(\varphi_{s_r} - cr\right)\left[r^{-1}\left(r\varphi_{o_r}\right)_r - \frac{n^2\varphi_o}{r^2}\right] - q_{s_r}\varphi_o = 0, \tag{4.116}$$

subject to the boundary condition $\varphi_o(a) = 0$. If we assume instability, i.e., $c = c_R + ic_I$, with $c_I \neq 0$, then multiplying (4.116) through by $r\varphi_o^*/\left(\varphi_{s_r} - cr\right)$, where φ_o^* is the complex conjugate of φ_o, and integrating the resulting expression with respect to r over the interval $(0,a)$ leads to the balance

$$\int_0^a \left\{|\varphi_{o_r}|^2 + \frac{n^2}{r^2}|\varphi_o|^2 + \frac{q_{s_r}\left(\varphi_{s_r} - c^*r\right)}{|\varphi_{s_r} - cr|^2}|\varphi_o|^2\right\} r\,dr = 0,$$

where we have integrated by parts once exploiting the boundary condition. The imaginary part of this balance is

$$c_I\int_0^a \frac{r^2 q_{s_r}|\varphi_o|^2}{|\varphi_{s_r} - cr|^2}dr = 0.$$

Thus if instability occurs, i.e., $c_I \neq 0$, then there exists at least one point $r^* \in (0,a)$ for which $q_{s_r}(r^*) = 0$. Conversely, if there does not exist a point $r^* \in (0,a)$ for which $q_{s_r}(r^*) = 0$, then $c_I \equiv 0$. We have therefore proven

Theorem 4.32 *(Rayleigh's Stability Theorem for circular flows.) A necessary condition for the normal mode instability of the steady circular flow $\varphi_s = \varphi_s(r)$ is that there exists at least one point in the flow domain $r^* \in (0, a)$ for which $q_{s_r}(r^*) = 0$. Conversely, if there does not exist a point $r^* \in (0, a)$ for which $q_{s_r}(r^*) = 0$, then the flow is neutrally stable.*

We now show that Rayleigh's Theorem for circular flows can be derived by obtaining conditions which will establish that $\delta^2 \mathcal{L}(q_s)$ is definite.

Theorem 4.33 *The circular flow $\varphi_s = \varphi_s(r)$ is linearly stable in the sense of Liapunov with respect to the perturbation enstrophy norm*

$$\|\delta\varphi\|^2 = \iint_{\Omega} (\Delta\delta\psi)^2 r^2 dr d\theta, \qquad (4.117)$$

if $\mathcal{R}'(q_s)$ is definite.

Proof. If there exist strictly positive real numbers Γ_1 and Γ_2 for which

$$0 < \Gamma_1 \leq \mathcal{R}'(q_s) \leq \Gamma_2 < \infty, \qquad (4.118)$$

then it follows from (4.113) that

$$2\Gamma_1 \|\delta\varphi\|^2 \leq \delta^2 \mathcal{L}(q_s)$$

$$= \left[\delta^2 \mathcal{L}(q_s)\right]_{t=0} \leq 2\Gamma_2 \|\delta\widetilde{\varphi}\|^2,$$

where $\delta\widetilde{\varphi}(r, \theta) = \delta\varphi(r, \theta, 0)$. Thus, $\forall \varepsilon > 0 \ \|\delta\widetilde{\varphi}\| \leq \varepsilon (\Gamma_1/\Gamma_2)^{\frac{1}{2}} \implies \|\delta\varphi\| \leq \varepsilon \ \forall \ t \geq 0$ and linear stability is established if (4.118) holds.

On the other hand, if there exist strictly negative real numbers Γ_1 and Γ_2 for which

$$-\infty < \Gamma_1 \leq \mathcal{R}'(q_s) \leq \Gamma_2 < 0, \qquad (4.119)$$

then it follows, exploiting the invariance of $\delta^2 \mathcal{L}(q_s)$, that

$$\|\delta\varphi(r, \theta, t)\| \leq \left(\frac{\Gamma_1}{\Gamma_2}\right)^{\frac{1}{2}} \|\delta\widetilde{\varphi}(r, \theta)\|,$$

where $\delta\widetilde{\varphi}(r, \theta) = \delta\varphi(r, \theta, 0)$. Thus, $\|\delta\widetilde{\varphi}\| \leq \varepsilon (\Gamma_2/\Gamma_1)^{\frac{1}{2}} \implies \|\delta\varphi\| \leq \varepsilon \ \forall \varepsilon > 0$ and $t \geq 0$ and linear stability is established if (4.119) holds. ∎

The relation to Rayleigh's Theorem 4.32 is straightforward. If follows from

$$\mathcal{R}(q_s) = r,$$

that

$$\mathcal{R}'(q_s)\, q_{s_r} = 1 \Longleftrightarrow \mathcal{R}'(q_s) = \frac{1}{q_{s_r}}, \quad (q_{s_r} \neq 0).$$

Thus we see that the definiteness of $\mathcal{R}'(q_s)$ is equivalent to the definiteness of the radial gradient of the mean flow vorticity q_s. Thus, Theorem 4.34 implies that a sufficient condition for the linear stability of a circular flow is that the radial gradient of the mean flow vorticity does not have a zero in the flow domain. This is just Theorem 4.32 proved as a consequence of the Hamiltonian structure.

Nonlinear generalization of Rayleigh's Theorem for circular flows

The nonlinear generalization of Theorem 4.33 is straightforward. Following the argument presented for parallel shear flows, we introduce the functional $\mathcal{N}(q)$ given by

$$\mathcal{N}(q) = \mathcal{L}(q_s + q) - \mathcal{L}(q_s) - \delta\mathcal{L}(q_s), \tag{4.120}$$

where $\mathcal{L}(q)$ is given by (4.112) and where $\delta\mathcal{L}(q_s)$ is written in the form

$$\delta\mathcal{L}(q_s) = \iint\limits_{\Omega} \left\{ \mathcal{R}^2 (\Delta\varphi_s) - r^2 \right\} q\, r dr d\theta = 0. \tag{4.121}$$

The *total* stream function, given by

$$\psi(r,\theta,t) = \varphi_s(r) + \varphi(r,\theta,t),$$

is the solution to the nonlinear vorticity equation (4.89). The disturbance stream function, given by $\varphi(r,\theta,t)$, satisfies the boundary condition $\varphi(a,\theta,t) = 0$. As before, we remark that since $\mathcal{N}(q)$ is the sum of three individually invariant functionals, it follows $\mathcal{N}(q)$ is itself an invariant of the nonlinear vorticity equation. Hence

$$\mathcal{N}(q(r,\theta,t)) = \mathcal{N}(\tilde{q}(r,\theta)),$$

where $\tilde{q}(r,\theta) = q(r,\theta,0)$ and $\tilde{\varphi}(r,\theta) = \varphi(r,\theta,0)$.

Substitution of (4.112) into (4.120) yields

$$\mathcal{N}(q) = \iint\limits_{\Omega} \left\{ \int\limits_{\Delta\varphi}^{\Delta\varphi_s + \Delta\varphi} \mathcal{R}^2(\xi)\, d\xi - \mathcal{R}^2(\Delta\varphi_s)\, \Delta\varphi \right\} r dr d\theta. \tag{4.122}$$

If $\mathcal{N}(\varphi)$ is Taylor expanded about $\varphi = 0$, one finds

$$\mathcal{N}(q) = \frac{1}{2}\delta^2\mathcal{L}(q_s) + h.o.t.,$$

where it is understood that the variations δq in $\delta^2 \mathcal{L}(q_s)$ are replaced with the disturbance vorticity $q(r,\theta,t)$.

If $\mathcal{R}^2(q)$ satisfies

$$\alpha \leq \frac{d[\mathcal{R}^2(q)]}{dq} \leq \beta, \qquad (4.123)$$

for some real constants α and β for all $q \in \mathbb{R}$, then it follows from Lemma 4.19 that

$$\frac{\alpha (\Delta\varphi)^2}{2} \leq \int\limits_{\Delta\varphi}^{\Delta\varphi_s + \Delta\varphi} \mathcal{R}^2(\xi)\, d\xi - \mathcal{R}^2(\Delta\varphi_s)\, \Delta\varphi \leq \frac{\beta (\Delta\varphi)^2}{2}, \qquad (4.124)$$

for all disturbance vorticity $\Delta\varphi$.

Theorem 4.34 *If*

$$0 < \alpha \leq \frac{d[\mathcal{R}^2(q)]}{dq} \leq \beta < \infty, \qquad (4.125)$$

or if

$$-\infty < \alpha \leq \frac{d[\mathcal{R}^2(q)]}{dq} \leq \beta < 0, \qquad (4.126)$$

for all $q \in \mathbb{R}$, then the circular flow solution $\psi = \varphi_s(r)$ is nonlinearly stable in the sense of Liapunov with respect to the disturbance enstrophy norm

$$\|\varphi\|^2 = \iint\limits_{\Omega} (\Delta\varphi)^2\, r\, dr\, d\theta. \qquad (4.127)$$

Proof. The proof is very similar to the proof presented for Theorem 4.29 and thus we will be brief in our presentation. If (4.122) holds, it follows from (4.122), (4.124) and the invariance of $\mathcal{N}(\varphi)$ that

$$0 < \alpha \|\varphi\|^2 \leq 2\mathcal{N}(\varphi)$$

$$= 2\mathcal{N}(\widetilde{\varphi}) \leq \beta \|\widetilde{\varphi}\|^2 < \infty.$$

Thus for every $\varepsilon > 0 \;\exists\; \delta = \varepsilon (\alpha/\beta)^{\frac{1}{2}} > 0$ such that $\|\widetilde{\varphi}\| \leq \delta \Longrightarrow \|\varphi\| \leq \varepsilon \;\forall\; t \geq 0$ if (4.125) holds.

On the other hand if (4.126) holds, then

$$-\infty < \alpha \|\widetilde{\varphi}\|^2 \leq 2\mathcal{N}(\widetilde{q})$$

$$= 2\mathcal{N}(q) \leq \beta \|\varphi\|^2 < 0,$$

from which it follows that

$$\|\varphi\| \leq \left(\frac{\alpha}{\beta}\right)^{\frac{1}{2}} \|\widetilde{\varphi}\|.$$

Thus for every $\varepsilon > 0 \ \exists \ \delta = \varepsilon (\beta/\alpha)^{\frac{1}{2}} > 0$ such that $\|\widetilde{\varphi}\| \leq \delta \implies \|\varphi\| \leq \varepsilon \ \forall \ t \geq 0$ if (4.126) holds. ∎

Our work on circular flows has been restricted to those flows for which the function $\mathcal{R}(q)$ is defined. This may not always be the case. Vortex patches are solutions to the vorticity equation which correspond to isolated (i.e., compactly supported but not necessarily simply connected) regions of constant vorticity. For these types of solutions it is usually necessary to further refine the analysis developed in this subsection. The interested reader is referred to, for example, Wan & Pulvirenti (1985), Wan (1986), Flierl (1988), Ripa (1987) and Kloosterziel & Carnevale (1992) for stability analyses of vortex patch solutions.

This concludes our discussion of the Hamiltonian structure and related stability theory for the incompressible two dimensional Euler equations. We turn now to a generalization of the two dimensional vorticity equation which includes two additional important kinematic processes found in fluids.

4.8   Exercises

Exercise 4.1 *Working from (4.116) show that a necessary condition for the instability of a circular flow $\varphi_s = \varphi_s(r)$ in the domain (4.105) subject to the boundary condition (4.106) is that there exists $r^* \in (0, a)$ such that $q_{s_r}(r^*)\varphi_{s_r}(r^*) < 0$, or conversely, a sufficient condition for stability is that $q_{s_r}(r)\varphi_{s_r}(r) \geq 0$ for all $r \in (0, a)$. This is Fjortoft's Theorem applied to circular flows.*

Exercise 4.2 *Show that the minimum eigenvalue, denoted $\lambda_{\min}$, of the eigenvalue problem*

$$\Delta\phi + \lambda\phi = 0, \quad (x, y) \in \Omega,$$

$$\phi|_{\partial\Omega} = 0,$$

where Ω is the circular domain (4.105), is given by

$$\lambda_{\min} = \left(\frac{j_{0,1}}{a}\right)^2 > 0,$$

where $j_{0,1}$ is the first zero of the Bessel function $J_0(*)$, i.e., $J_0(j_{0,1}) = 0$.

Exercise 4.3 *Show that the only flows that can satisfy the stability conditions of Theorem 4.15 or 4.17 in the circular domain (4.105) with the boundary condition (4.106) are circular flows of the form $\varphi_s = \varphi_s(r)$.*

Chapter 5

The *Charney-Hasegawa-Mima* equation

One unsatisfactory feature of the two dimensional vorticity equation is that it is, of course, incapable of modelling any three dimensional effects such as vortex tube stretching/compression. The dilation and compression of vortex tubes and filaments is a fundamental characteristic of turbulent three dimensional fluid flows. Another feature that the two dimensional vorticity equation (3.8) does not include, in the absence of a mean flow, is a term corresponding to a background vorticity gradient. The most important consequence of this property is that there are no nontrivial linear plane wave solutions to (3.8).

The simplest generalization of the two dimensional vorticity equation which includes these effects is the *Charney-Hasegawa-Mima (CHM)* equation which can be written in the nondimensional form

$$(\Delta - 1)\,\psi_t + \psi_x + \partial\,(\psi, \Delta\psi) = 0. \tag{5.1}$$

The terms proportional to ψ_t and ψ_x in (5.1) model, respectively, vortex tube stretching and a background vorticity gradient. The *CHM* equation may be considered a canonical equation for quasi-two dimensional flow in the presence of a relatively constant background vorticity gradient. It is usually obtained via a formal asymptotic reduction of the relevant governing equations. For example, Hasegawa & Mima (1978) have shown that (5.1) describes the leading order evolution of the electrostatic potential for a weakly magnetized nonuniform plasma. In the context of geophysical fluid dynamics, equation (5.1) describes the low wavenumber and frequency evolution of the leading order dynamic pressure for a rapidly rotating fluid of finite thickness (Charney, 1947; see also LeBlond & Mysak, 1978 or Pedlosky, 1987).

There are, however, some symmetry properties which the two dimensional vorticity equation possesses but the *CHM* equation does not retain. One property is rotational invariance. The two dimensional vorticity equation is invariant for arbitrary rotations of the coordinates (x, y) but the *CHM* equation is not. The loss of rotational invariance can be attributed to the presence of the background vorticity term ψ_x in (5.1).

Another property that is lost is Galilean invariance. The two dimensional vorticity equation is invariant for an arbitrary Galilean transformation of the form

$$\psi(x, y, t) \longrightarrow \psi(x - ct, y, t) + cy,$$

for arbitrary c but the *CHM* equation is not. The loss of Galilean invariance can be traced to the presence of the vortex stretching term ψ_t in (5.1)

The *CHM* equation has dispersive linear plane wave solutions. If the nonlinear term $\partial(\psi, \Delta\psi)$ is neglected in (5.1), then

$$(\Delta - 1)\psi_t + \psi_x \approx 0. \tag{5.2}$$

This equation has a plane wave solution of the form

$$\psi = Real\{a\exp[i(kx + ly - \omega t)]\}, \tag{5.3}$$

provided the dispersion relationship

$$\omega = -\frac{k}{k^2 + l^2 + 1}, \tag{5.4}$$

holds.

These solutions, which play an important role in the large scale evolution of planetary atmospheres and oceans, are call *Rossby* waves (Rossby, 1939). Although the *CHM* equation was explicitly linearized to obtain the Rossby wave dispersion relation, it is easy to verify that a *single* Rossby wave is an exact solution of the fully nonlinear *CHM* equation (i.e., $\partial(\psi, \Delta\psi) = 0$ for ψ given by (5.3) and (5.4)). However, a linear superposition of Rossby waves is, in general, *not* a solution of the full nonlinear *CHM* equation.

The *CHM* equation (5.1) also has steadily travelling dipole vortex solutions (Stern, 1975; Larichev & Reznik, 1976) with compact support (in the sense that ψ decays to zero exponentially rapidly at infinity). We will describe these solutions in Section 5.5.

In the next section we give a derivation of the *CHM* equation and the appropriate boundary conditions in the context of the shallow-water equations for a differentially rotating fluid of finite depth. This system of equations provides a useful model for many of the most important large scale dynamical processes observed in planetary atmospheres and oceans. The reader who is already familiar with this derivation may wish to skip over the next section and proceed directly to the presentation of the Hamiltonian structure.

5.1   A derivation of the *CHM* equation

The *dimensional* shallow water equations for the differentially rotating fluid depicted in Fig. 5.1 can be written in the form

$$\left(\frac{\partial}{\partial t^*} + \mathbf{u}^* \cdot \nabla^*\right) u^* - (f_o + \beta y^*)v^* + g\eta_{x^*}^* = 0, \tag{5.5}$$

$$\left(\frac{\partial}{\partial t^*} + \mathbf{u}^* \cdot \nabla^*\right) v^* + (f_o + \beta y^*)u^* + g\eta_{y^*}^* = 0, \tag{5.6}$$

$$\eta_{t^*}^* + \nabla^* \cdot [\mathbf{u}^* (H + \eta^*)] = 0, \tag{5.7}$$

where the velocity field is given by

$$\mathbf{u}^* = (u, v),$$

and where the free surface elevation is given by $\eta^* (x^*, y^*, t^*)$, and where

$$\nabla^* = \left(\frac{\partial}{\partial x^*}, \frac{\partial}{\partial y^*}\right).$$

The constant mean fluid thickness is given by H.

The *Coriolis* "force" may be written in the vector form

$$2\Omega \times \mathbf{u}^* \equiv (f_o + \beta y^*)\mathbf{e}_3 \times \mathbf{u}^*,$$

where the angular frequency vector is given by

$$\Omega \equiv \frac{1}{2}(f_o + \beta y^*)\mathbf{e}_3,$$

(see Fig. 5.1). Observe that the magnitude of the angular frequency vector is a linear function of y^*. The parameter β will be assumed to be positive, implying that the angular frequency of the fluid increases linearly in the positive y-coordinate direction. In large scale atmospheric and oceanographic dynamics, it is the local radially outward component of the *earth's* angular frequency vector (which is oriented along the earth's spin axis in the northward direction) which is dynamically relevant. (For a complete discussion of these ideas see LeBlond & Mysak, 1978 or Pedlosky, 1987.)

However, the magnitude of the local radially outward component of the earth's angular frequency vector is a function of latitude (but not of longitude). Thus, if we choose a local Cartesian coordinate system in which the x^*, y^*, and z^* coordinate axes point, respectively, in the eastward, northward and upward directions, then the radially outward component of the earth's angular frequency vector will be a function of y^*. If the north-south length scale associated with the dynamics is not too large (but still large enough for the underlying assumptions implicit in shallow-water theory to be valid), then the dependence of the radially outward component of the earth's angular frequency vector on y^* can be reasonably well approximated with a linear function in y^*.

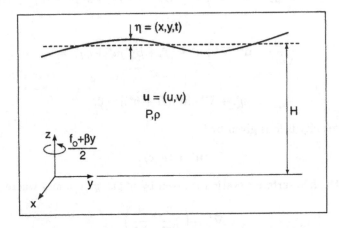

Figure 5.1. Geometry of the configuration used to derive the *CHM* equation.

It is this approximation which gives rise to

$$\Omega \equiv \frac{1}{2}(f_o + \beta y^*)\mathbf{e}_3,$$

where $f_o \equiv 2\Omega_E \sin{(\theta_o)}$ and $\beta \equiv 2\Omega_E \cos{(\theta_o)}/R$, where Ω_E is the magnitude of the earth's angular frequency vector ($\Omega_E = 2\pi \ rads/day$), θ_o is the latitude at which the origin of the local (x^*, y^*, z^*) Cartesian coordinate system is placed, and R is the equatorial radius of the earth ($R \simeq 6.4 \times 10^6 m$). For midlatitudes ($\theta_o \simeq 45^\circ$), $f_o \simeq 10^{-4} \sec^{-1}$ and $\beta \simeq 10^{-11} m^{-1} \sec^{-1}$. The absolute size of these terms should not lead the reader to think that they play a secondary role in the evolution of the flow. It turns out that the terms proportional to f_o and the pressure gradient are the most important terms in the momentum equations, and the term proportional to β allows for the possibility of Rossby waves.

Let us *nondimensionalize* the shallow-water equations as follows:

$$(x^*, y^*) = R_*(x, y), \ t^* = (R_*\beta)^{-1} t, \ R_* \equiv \frac{\sqrt{gH}}{f_o}, \tag{5.8}$$

$$(u^*, v^*) = \beta R_*^2 (u, v), \ \eta^* = \frac{H\beta R_*}{f_o}\eta, \tag{5.9}$$

where the unasterisked variables correspond to the nondimensional quantities. Substitution of (5.8) and (5.9) into (5.5), (5.6) and (5.7) leads to

$$\varepsilon \left(\frac{\partial}{\partial t} + \mathbf{u} \cdot \nabla \right) u - (1 + \varepsilon y)v + \eta_x = 0, \tag{5.10}$$

$$\varepsilon \left(\frac{\partial}{\partial t} + \mathbf{u} \cdot \nabla \right) v + (1 + \varepsilon y)u + \eta_y = 0, \tag{5.11}$$

$$\varepsilon \eta_t + \nabla \cdot [\mathbf{u} (1 + \varepsilon \eta)] = 0, \tag{5.12}$$

where

$$\varepsilon \equiv \frac{\beta R_*}{f_o}.$$

The nondimensional parameter ε is called the *Rossby number*. In the context of (5.10) and (5.11) its physical interpretation is that it is a measure of the order of magnitude of the convective derivative and the β-effect to the presumably dominant balance between the leading order Coriolis term and the pressure gradient. In (5.12), ε can be interpreted as a measure of the nondivergence of the horizontal velocity field.

For large scale atmospheric (where the horizontal length and velocity scales are, respectively, on the order of 10^3 *km* and 10 *m/s*) and oceanic motions (where the horizontal length and velocity scales are, respectively, on the order of 10^2 *km* and 10 *cm/s*), it is the case that $\varepsilon \simeq 10^{-1}$. This fact motivates attempting to solve (5.10), (5.11) and (5.12) by constructing a straightforward asymptotic expansion of the form

$$[u (x, y, t; \varepsilon), v (x, y, t; \varepsilon), \eta (x, y, t; \varepsilon)] \simeq$$

$$[u^{(0)} (x, y, t), v^{(0)} (x, y, t), \eta^{(0)} (x, y, t)]$$

$$+ \varepsilon [u^{(1)} (x, y, t), v^{(1)} (x, y, t), \eta^{(1)} (x, y, t)] + O (\varepsilon^2). \tag{5.13}$$

The $O (1)$ problem is given by

$$v^{(0)} = \eta_x^{(0)}, \tag{5.14}$$

$$u^{(0)} = -\eta_y^{(0)}, \tag{5.15}$$

$$u_x^{(0)} + v_y^{(0)} = 0. \tag{5.16}$$

The expressions (5.14) and (5.15) imply that, to leading order, the velocity is determined by a balance between the Coriolis term and the pressure gradient. This balance is referred to as the *geostrophic balance*. Another to note is that the leading order pressure forms a stream function for the leading order velocity field. Thus, to leading order, the flow follows lines of constant pressure.

The problem with the $O(1)$ equations is that does not form a closed set of equations. Observe that (5.16) is trivially satisfied for *every* sufficiently smooth $\eta^{(0)}(x, y, t)$. To uniquely determine the $O(1)$ solution it is necessary to examine the $O(\varepsilon)$ equations, given by

$$\left(\frac{\partial}{\partial t} + \mathbf{u}^{(0)} \cdot \nabla\right) u^{(0)} - v^{(1)} - y v^{(0)} + \eta_x^{(1)} = 0, \tag{5.17}$$

$$\left(\frac{\partial}{\partial t} + \mathbf{u}^{(0)} \cdot \nabla\right) v^{(0)} + u^{(1)} + y u^{(0)} + \eta_y^{(1)} = 0, \tag{5.18}$$

$$\eta_t^{(0)} + \nabla \cdot \left[\mathbf{u}^{(1)} + \mathbf{u}^{(0)} \eta^{(0)}\right] = 0. \tag{5.19}$$

The relevant equation for $\eta^{(0)}$ is obtained by constructing the vorticity equation associated with the $O(\varepsilon)$ problem. If we form

$$\partial_x(5.18) - \partial_y(5.17),$$

it follows that

$$\left(\frac{\partial}{\partial t} + \mathbf{u}^{(0)} \cdot \nabla\right) \left(v_x^{(0)} - u_y^{(0)}\right) + v^{(0)} + \left(u_x^{(1)} + v_y^{(1)}\right) = 0, \tag{5.20}$$

where we have used (5.16) to simplify the derivatives associated with the advective terms in the convective derivative. From (5.19) we find

$$\nabla \cdot \mathbf{u}^{(1)} = -\left(\frac{\partial}{\partial t} + \mathbf{u}^{(0)} \cdot \nabla\right) \eta^{(0)}, \tag{5.21}$$

which substituted into (5.20) gives

$$\left(\frac{\partial}{\partial t} + \mathbf{u}^{(0)} \cdot \nabla\right) \left(v_x^{(0)} - u_y^{(0)} - \eta^{(0)}\right) + v^{(0)} = 0. \tag{5.22}$$

If the geostrophic relations (5.14) and (5.15) are substituted into (5.22), we obtain the *CHM* equation in the form

$$(\Delta - 1)\eta_t^{(0)} + \eta_x^{(0)} + \partial\left(\eta^{(0)}, \Delta\eta^{(0)}\right) = 0. \tag{5.23}$$

In the context of large scale ocean and atmospheric dynamics, (5.23) is usually referred to as the *quasi-geostrophic potential vorticity equation* associated with the shallow-water equations. Observe that (5.23) can be written in the form

$$\left(\frac{\partial}{\partial t} + \mathbf{u}_g^{(0)} \cdot \nabla\right) \left(\triangle \eta^{(0)} - \eta^{(0)} + y\right) = 0, \tag{5.24}$$

where

$$\mathbf{u}_g^{(0)} \equiv \mathbf{e}_3 \times \nabla \eta^{(0)}.$$

Thus, in the context of large scale ocean and atmospheric dynamics, (5.24) is the statement that the quasi-geostrophic potential vorticity, given by

$$\triangle \eta^{(0)} - \eta^{(0)} + y,$$

is conserved following the geostrophic motion. The quasi-geostrophic potential vorticity is composed of three terms. The first term in the potential vorticity, given by

$$\triangle \eta^{(0)} = \mathbf{e}_3 \cdot \left[\nabla \times \mathbf{u}_g^{(0)}\right],$$

is the vertical component of the curl of the velocity field, which sometimes referred to as the *relative vorticity*. The second term, given by $-\eta^{(0)}$, is a vortex-tube stretching term associated with the free surface. The third term, given by y, is the vorticity associated with the differential rotation of the fluid, which is sometimes referred to as the planetary vorticity gradient.

Alternatively, (5.24) may be interpreted as the leading order term associated with *Ertel's Theorem* (e.g., Pedlosky, 1987) for the shallow-water equations. Ertel's Theorem, for the nondimensional shallow-water model (5.10), (5.11) and (5.12), is the statement that

$$\left(\frac{\partial}{\partial t} + \mathbf{u} \cdot \nabla\right) \left[\frac{v_x - u_y + 1 + \varepsilon y}{1 + \varepsilon \eta}\right] = 0, \tag{5.25}$$

holds for (5.10), (5.11) and (5.12). (It is left as an exercise for the reader to derive (5.25) directly from (5.10), (5.11) and (5.12).) The quantity

$$\frac{v_x - u_y + 1 + \varepsilon y}{1 + \varepsilon \eta},$$

is called the shallow-water potential vorticity. Obviously, Ertel's Theorem is the statement that the potential vorticity is conserved following the motion. If the asymptotic expansion (5.13) is inserted into (5.25), the (nontrivial) leading order expression that is obtained is just the *CHM* equation (5.24).

Boundary Conditions

The boundary conditions for the *CHM* equation are identical to those assumed for the two dimensional vorticity equation (3.8). It will be assumed that the flow occurs on the domain $\Omega \subseteq \mathbb{R}^2$. The boundary of Ω, if it exists, will be denoted by $\partial\Omega$ and it will be assumed to be the union of a finite number of smooth simply connected curves $\{\partial\Omega_i\}_{i=1}^n$ so that $\partial\Omega = \cup_{i=1}^n \partial\Omega_i$. If the asymptotic expansion (5.13) is inserted into the no-slip condition (3.14) it follows that the leading order boundary condition is given by

$$\left[\mathbf{n} \cdot \mathbf{u}^{(0)}\right]_{\partial\Omega} = 0, \tag{5.26}$$

which implies that

$$\eta^{(0)} = \lambda_i, \tag{5.27}$$

(where the λ_i are constants) on $\partial\Omega_i$ for $i = 1, ..., n$. As in our discussion of the boundary conditions for the two dimensional vorticity equation, if the domain is bounded and simply connected then we may, without loss of generality, assume $\eta^{(0)} = \lambda = 0$ on $\partial\Omega$. Throughout our discussion of the *CHM* equation, we will assume that the boundary conditions are given by (5.27). If other conditions are required, they will be specified as the need arises.

Existence and uniqueness of solutions

As in our discussion of the two dimensional vorticity equation, it is beyond the scope of this book to give a proper account of the mathematical existence, uniqueness and regularity theory of solutions to the Cauchy problem associated with the *CHM* equation given by

$$(\Delta - 1)\psi_t + \psi_x + \partial(\psi, \Delta\psi) = 0,$$

for $(x, y, t) \in \Omega \times [0, \infty)$, with the boundary conditions

$$[\mathbf{n} \times \nabla\psi]_{\partial\Omega} = 0, \quad \forall \quad t \geq 0,$$

and the initial conditions

$$\psi(x, y, 0) = \psi_0(x, y),$$

$$[\mathbf{n} \times \nabla\psi_0]_{\partial\Omega} = 0.$$

Here, we will simply point out that the existence of *weak* solutions and the uniqueness of *classical* or *strong* solutions to the Cauchy problem for a generalized

CHM equation appropriate for a density-stratified fluid was first established by Dutton (1974). The existence of classical solutions to the Cauchy problem for the density-stratified *CHM* equation examined by Dutton was subsequently proved by Bennett & Kloeden (1980). In addition, Bennett & Kloeden (1981, 1982) went on to develop the mathematical existence, uniqueness and regularity theory of classical solutions to the Cauchy problem associated with *CHM*-like equations in other contexts. The reader who is interested in pursuing these interesting issues further is referred to these papers and the references contained therein.

5.2 Hamiltonian structure

It is not surprising that since the *CHM* equation differs from the two dimensional vorticity equation (3.8) in containing only two additional terms, its Hamiltonian structure is very similar. As a result we will be relatively brief in our presentation and many of the proofs will be deleted. Weinstein (1983) gave the following Hamiltonian formulation of the *CHM* equation. We also remark that while there exists a variational principle for the *CHM* equation in terms of Lagrangian variables (Virasoro, 1981), one does not exist in terms of Eulerian variables unless a Clebsch variable transformation is introduced.

Theorem 5.1 *The CHM equation (5.1) can be cast into the scalar Hamiltonian form*

$$q_t = J\delta H/\delta q, \tag{5.28}$$

where the Hamiltonian $H(q)$ and J are given, respectively, by

$$H = \frac{1}{2}\iint_{\Omega} \nabla\psi \cdot \nabla\psi + \psi^2 dx dy - \sum_{i=1}^{n} \lambda_i \oint_{\partial\Omega_i} \mathbf{n}_i \cdot \nabla\psi ds, \tag{5.29}$$

$$J(*) = -\partial(q, *), \tag{5.30}$$

and where the variable q is the total vorticity given by

$$q = \Delta\psi - \psi + y. \tag{5.31}$$

Proof. The proof of this theorem is identical to that presented for Theorem 3.1 and left as an exercise for the reader. ∎

The Hamiltonian (5.29) is comprised of two individually invariant terms. The first is the total energy, given by

$$E_{total} \equiv \frac{1}{2}\iint_{\Omega} \nabla\psi \cdot \nabla\psi + \psi^2 dx dy.$$

The first term in the integrand, given by $\nabla \psi \cdot \nabla \psi$, is the contribution associated with the *kinetic* energy. The second term in the integrand, given by ψ^2, is the contribution associated with the *potential* energy. It is easy to verify that E_{total} is conserved. The second term in the Hamiltonian may be viewed as a weighted sum of the circulation integrals

$$\Gamma_i \equiv \oint_{\partial \Omega_i} \mathbf{n}_i \cdot \nabla \psi ds, \ i = 1, ..., n.$$

Each of these circulations integrals is, of course, itself invariant due to *Kelvin's Circulation Theorem.*

Poisson Bracket Formulation

The Poisson bracket associated with the above Hamiltonian formulation for the *CHM* equation is given by

$$[F, G] = -\left\langle \frac{\delta F}{\delta q} , \partial(q, \frac{\delta G}{\delta q}) \right\rangle, \tag{5.32}$$

where F and G are arbitrary functionals with respect to q. Consequently, the evolution of an arbitrary functional $F(q)$ will be determined by

$$\frac{\partial F}{\partial t} = [F, H]$$

$$= -\left\langle \frac{\delta F}{\delta q} , \partial(q, \frac{\delta H}{\delta q}) \right\rangle = \iint_{\Omega} \frac{\delta F}{\delta q} q_t \ dxdy. \tag{5.33}$$

In particular, therefore, the *CHM* equation can be re-written in the form

$$q_t = [q, H], \tag{5.34}$$

provided we interpret q as the functional

$$q(x, y, t) = \iint_{\Omega'} q(x', y', t) \delta(\mathbf{x} - \mathbf{x}') dx'dy', \tag{5.35}$$

where Ω' is the domain Ω parametrized by (x', y').

It is, of course, necessary to explicitly check that the bracket (5.32) does indeed satisfy the algebraic properties (3.20) through (3.24) of the general definition of a Poisson bracket. This is left as an exercise for the reader. We note, however, as in our proof of Theorem 3.1, it is necessary to restrict the set of allowed functionals to those whose variational derivatives evaluated on the boundary can be explicitly written as a function of q. That is, if F is an allowed functional, then

$$\left[\frac{\delta F}{\delta q} \right]_{\partial \Omega} = f(q).$$

Casimir invariants

The Casimir invariants associated with this Hamiltonian structure, denoted C, are the solutions of

$$J \frac{\delta C}{\delta q} \equiv -\partial(q, \frac{\delta C}{\delta q}) = 0. \tag{5.36}$$

This can be integrated to give

$$C(q) \equiv \iint_\Omega \Phi(q) - \Phi(y) \, dx dy, \tag{5.37}$$

where the Casimir density $\Phi(*)$ is an arbitrary function of its argument. The second term in the integrand is not needed if the domain is bounded and must removed from the integral if the domain is unbounded and $q \longrightarrow 0$ (i.e., $\Delta \psi - \psi \longrightarrow -y$) as $x^2 + y^2 \longrightarrow \infty$.

It necessarily follows that C is conserved, i.e., $[H, C] = 0$. This can be verified directly as follows. From (5.37) we find

$$\frac{dC}{dt} = \iint_\Omega \Phi'(q) \, q_t \, dx dy,$$

where

$$\Phi'(q) = \frac{d\Phi(q)}{dq}.$$

Since $q_t = -\partial(\psi, q)$, this can be re-written in the form

$$\frac{dC}{dt} = -\iint_\Omega \Phi'(q) \, \partial(\psi, q) \, dx dy$$

$$= -\iint_\Omega \partial(\psi, \Phi(q)) \, dx dy$$

$$= -\iint_\Omega \nabla \cdot [\Phi(q) \, \mathbf{e}_3 \times \nabla \psi] \, dx dy$$

$$= \sum_{i=1}^n \oint_{\partial \Omega_i} \frac{\partial \psi}{\partial s} \Phi(q) \, ds = 0.$$

Application of Noether's Theorem

The general application of Noether's Theorem to symmetries in the Hamiltonian structure for the *CHM* equation is similar to the theory outlined for the two dimensional vorticity equation and is not repeated here. The only symmetry for which we wish to explicitly determine the corresponding time invariant functional is the invariance of the Hamiltonian structure to arbitrary translations in x. The conserved functional associated with this symmetry is, of course, Kelvin's Impulse in the x-direction (Lamb, 1932) associated with the *CHM* equation.

As in our corresponding discussion of the two dimensional vorticity equation, let us consider the periodic channel domain

$$\Omega = \{(x,y) \mid -x_o < x < x_o, \ 0 < y < L\}, \tag{5.38}$$

with the boundary conditions $\psi = \psi_1$ and ψ_2 on $y = 0$ and L, respectively, and where it is assumed that ψ is smoothly periodic at $x = \pm x_o$. The Hamiltonian for this domain can be written in the form

$$H = \frac{1}{2} \iint\limits_{\Omega} \nabla\psi \cdot \nabla\psi + \psi^2 dx dy$$

$$-\psi_2 \int\limits_{-x_o}^{x_o} \psi_y(x,L,t)\,dx + \psi_1 \int\limits_{-x_o}^{x_o} \psi_y(x,0,t)\,dx. \tag{5.39}$$

It is straightforward to verify that H and J are invariant under the transformation $x \longrightarrow x + \tau$, for arbitrary τ. Hence, Noether's Theorem implies the existence of a conserved function, denoted $M(q)$, satisfying

$$J\frac{\delta M}{\delta q} = -q_x. \tag{5.40}$$

Substituting in for J into (5.40) gives

$$\partial\left(q, \frac{\delta M}{\delta q}\right) = q_x,$$

which can be integrated to give the general expression

$$M(q) = \iint\limits_{\Omega} y\,(q - y)\,dx dy. \tag{5.41}$$

In a bounded domain, such as Ω given in (5.38), the $-y^2$ term in the integrand of (5.41) is not necessary and can be omitted. However, if the domain is unbounded and ψ has, for example, compact support, then this term is necessary if $M(q)$ is to exist (i.e., be finite).

Although Noether's Theorem guarantees that $M(q)$ is invariant in time, let us verify this directly. From (5.41) we have

$$\frac{dM}{dt} = \iint_{\Omega} y q_t \, dx dy = -\iint_{\Omega} y \partial(\psi, q) \, dx dy$$

$$= -\iint_{\Omega} y \nabla \cdot [q \mathbf{e}_3 \times \nabla \psi] \, dx dy$$

$$= \iint_{\Omega} \mathbf{e}_2 \cdot [q \mathbf{e}_3 \times \nabla \psi] \, dx dy$$

$$= \iint_{\Omega} q \psi_x dx dy = -\iint_{\Omega} q_x \psi dx dy$$

$$= -\iint_{\Omega} (\Delta \psi_x - \psi_x) \psi dx dy$$

$$= \frac{1}{2} \iint_{\Omega} (\nabla \psi \cdot \nabla \psi)_x \, dx dy = 0,$$

where we have repeatedly integrated by parts exploiting the boundary and periodicity conditions.

5.3 Steady solutions

General theory

Suppose that $\psi = \varphi_s(x, y)$ is a steady solution to the *CHM* equation (5.1). Substituting into (5.1) gives

$$\partial(\varphi_s, \Delta \varphi_s - \varphi_s + y) = 0. \tag{5.42}$$

As in the two dimensional vorticity equation, we see that arbitrary steady solutions are described by the fact that the Jacobian between the stream function and the (total) vorticity is zero. It follows from (5.42) that

$$\varphi_s = \Phi(\Delta \varphi_s - \varphi_s + y), \tag{5.43}$$

where $\Phi(*)$ is a possibly nonanalytic function of its argument. Thus we have proved

Lemma 5.2 *General steady solutions of the CHM equation satisfy the condition that the stream function ψ can be written as a function of the total vorticity $\Delta\psi - \psi + y$ in the form (5.43).*

Variational Principles

General steady solutions to the *CHM* equation can be characterized with the variational principle

Theorem 5.3 *Suppose that $\psi = \varphi_s(x, y)$ is an arbitrary steady solution to the CHM equation given by (5.43). Then $\varphi_s(x, y)$ satisfies the first order necessary conditions for an extremal of the constrained Hamiltonian*

$$\mathcal{H}(q) = H(q) + \iint\limits_{\Omega} \left\{ \int\limits_{y}^{q} \Phi(\xi)\, d\xi \right\}\, dxdy. \tag{5.44}$$

Proof. The proof is straightforward. The first variation of $\mathcal{H}(q)$ is given by

$$\delta\mathcal{H}(q) = \iint\limits_{\Omega} [\Phi(q) - \psi]\, \delta q\, dxdy, \tag{5.45}$$

where we integrated by parts once and used the boundary conditions. It follows from (5.45) that

$$\delta\mathcal{H}(q_s) = \iint\limits_{\Omega} [\Phi(q_s) - \varphi_s]\, \delta q\, dxdy = 0,$$

where

$$q_s \equiv \Delta\varphi_s - \varphi_s + y,$$

and (5.43) has been used. ∎

    An alternate variational principle, based on a constrained x-direction impulse functional similar to that presented in Theorem 4.26, can be given for parallel shear flow solutions of the form $\varphi_s = \varphi_s(y)$. One can verify that $\psi = \varphi_s(y)$ is an exact solution by direct substitution in (5.1) and noting that

$$\frac{\partial\varphi_s}{\partial t} \equiv 0,$$

$$\frac{\partial\varphi_s}{\partial x} \equiv 0,$$

$$\partial\left(\varphi_s, \Delta\varphi_s\right) \equiv 0.$$

These solutions correspond to two dimensional flows which are oriented perpendicular to the direction of the background vorticity gradient. For example, in the context of geophysical fluid dynamics these solutions correspond to flows which are oriented in an east-west configuration. They can be thought of as models, for example, of the averaged large scale winds that occur in the midlatitudes. Note that due to the presence of the background vorticity term ψ_x in (5.1), there are no steady solutions of the form $\varphi_s = \varphi_s(x)$.

To make our discussion concrete, let us assume the that flow takes place in the periodic channel domain given by

$$\Omega = \{(x, y) \, |x| < x_o, \, 0 < y < L\}, \tag{5.46}$$

with the boundary conditions

$$\psi = \psi_1 \text{ and } \psi_2 \text{ on } y = 0 \text{ and } L, \tag{5.47}$$

respectively, and where it is assumed that ψ is smoothly periodic at $x = \pm x_o$.

Consider the parallel shear flow solution $\psi = \varphi_s(y)$ where $\varphi_s = \psi_1$ and ψ_2 on $y = 0$ and L, respectively. Let $\mathcal{Y}(q_s)$ be the inverse function associated with $q_s = q_s(y)$, i.e.,

$$\mathcal{Y}\left(q_s(y)\right) = y.$$

Let $\mathcal{L}(q)$ be the constrained x-direction impulse functional

$$\mathcal{L}(q) = \iint_\Omega \left\{ \int_y^q \mathcal{Y}(\xi)\, d\xi - y(q - y) \right\} dx\, dy$$

$$= \iint_\Omega \left\{ \int_y^{\Delta\psi - \psi + y} \mathcal{Y}(\xi)\, d\xi - y(\Delta\psi - \psi) \right\} dx\, dy.$$

The first variation of $\mathcal{L}(q)$ is given by

$$\delta\mathcal{L}(q) = \iint_\Omega \left\{ \mathcal{Y}(q) - y \right\} \delta q \, dx\, dy.$$

From which it follows that $\delta\mathcal{L}(q_s) = 0$ since $\mathcal{Y}(q_s(y)) = y$. We have therefore established the variational principle

Theorem 5.4 *Let* $\psi = \varphi_s(y)$ *be a parallel shear flow solution of the CHM equation (5.1) in the periodic channel domain (5.46) satisfying the boundary conditions (5.47) and the smooth periodicity conditions at* $x = \pm x_o$, *then* $\delta\mathcal{L}(q_s) = 0$ *for* $\mathcal{L}(q)$ *given by*

$$\mathcal{L}(q) = \iint_\Omega \left\{ \int_y^q \mathcal{Y}(\xi)\, d\xi - y\,(q-y) \right\} dx\,dy, \tag{5.48}$$

where $\mathcal{Y}(q)$ *is the inverse function associated with*

$$q = q_s(y) \equiv \varphi_{s_{yy}}(y) - \varphi_s(y) + y.$$

5.4   Stability of steady solutions

The stability theory for steady solutions to the *CHM* equation is very similar to that presented for the two dimensional vorticity equation. As a result, we will be brief in our presentation of the main results. Our discussion, while abbreviated, will follow a format similar in outline to that presented in Chapter 4.

General linear stability equation

Suppose we assume a solution to the *CHM* equation in the form

$$\psi(x,y,t) = \varphi_s(x,y) + \varphi(x,y,t),$$

where φ_s is the steady solution given by (5.43). Substituting this decomposition into (5.1) and neglecting all quadratic terms in the perturbation stream function $\varphi(x,y,t)$ leads to

$$(\Delta - 1)\varphi_t + \partial(\varphi_s, \Delta\varphi - \varphi) + \partial(\varphi, \Delta\varphi_s - \varphi_s + y) = 0. \tag{5.49}$$

If (5.43) is substituted into this equation, the result can be written in the form

$$(\Delta - 1)\varphi_t + \partial(q_s, \Phi_s'\Delta\varphi - [\Phi_s' + 1]\varphi) = 0, \tag{5.50}$$

with

$$\Phi_s' \equiv \left[\frac{d\Phi(q)}{dq} \right]_{q=q_s}. \tag{5.51}$$

where $q_s = \Delta\varphi_s - \varphi_s + y$. Equation (5.50) is the general linear stability equation associated with an arbitrary steady solution $\varphi_s(x,y)$ of the *CHM* equation.

   If we assume that

$$\psi|_{\partial\Omega_i} = \varphi_s|_{\partial\Omega_i} = \lambda_i, \tag{5.52}$$

for $i = 1, ..., n$, then the appropriate perturbation boundary conditions are

$$\psi|_{\partial \Omega_i} = 0, \tag{5.53}$$

for $i = 1, ..., n$. If other boundary or periodicity conditions are required, they will specified at that time.

Linear stability equation for a parallel shear flow

The general linear stability equation (5.50) can be considerably simplified for a parallel shear flow solution for the form $\varphi_s = \varphi_s(y)$. For a steady solution of this form, (5.50) reduces to

$$(\Delta - 1)\,\varphi_t + \left(U_{s_{yy}} - U_s - 1\right) \Phi'_s \left(\Delta \varphi - \varphi\right)_x$$

$$- \left(U_{s_{yy}} - U_s - 1\right) \varphi_x = 0, \tag{5.54}$$

where

$$U_s(y) \equiv -\frac{d\varphi_s(y)}{dy}. \tag{5.55}$$

However, if (5.43) is differentiated with respect to y, it follows that

$$\Phi'_s = \frac{U_s}{U_{s_{yy}} - U_s - 1}.$$

If this expression is substituted into (5.54), the resulting equation can be arranged into the form

$$[\partial_t + U_s \partial_x] (\Delta \varphi - \varphi) + \left(1 - U_{s_{yy}} + U_s\right) \varphi_x = 0,$$

or equivalently

$$[\partial_t + U_s \partial_x] (\Delta \varphi - \varphi) + \frac{dq_s}{dy} \varphi_x = 0, \tag{5.56}$$

where $q_s = \varphi_{s_{yy}} - \varphi_s + y$ for the parallel shear flow (5.55). This is the general linear stability equation for a parallel shear flow of the form $\varphi_s = \varphi_s(y)$. Note that (5.56) differs from (4.25) in that it includes the additional terms associated with vortex tube stretching and differential rotation.

Normal mode equations for parallel shear flows

The normal mode equations associated with (5.56) are obtained by assuming a solution of the form

$$\varphi\left(x,y,t\right) = \varphi_o(y) \exp\left[ik\left(x - ct\right)\right] + c.c., \tag{5.57}$$

where $k = \dfrac{n\pi}{x_o}$, for $n = 0, 1, 2, \ldots$ (due to the periodicity at $x = \pm x_o$), with k the x-direction wavenumber, and the complex-valued phase speed $c = c_R + ic_I$.

Substitution of (5.57) into (5.56) leads to

$$\left(U_s - c\right)\left(\varphi_{o_{yy}} - k^2\varphi_o - \varphi_o\right) + \left(1 - U_{s_{yy}} + U_s\right)\varphi_o = 0, \tag{5.58}$$

$$\varphi_o\left(0\right) = \varphi_o\left(L\right) = 0. \tag{5.59}$$

This two-point homogeneous boundary value problem is an eigenvalue problem. In general, it will have a nontrivial solution only if certain special relationships between the complex-valued phase speed c and the x-direction wavenumber k hold. Since (5.58) and (5.59) are invariant under the transformation $k \to -k$, we may, without loss of generality, assume $k \geq 0$. In addition, if (φ_o, c) is a solution, to (5.58) and (5.59), for a given wavenumber, then so is (φ_o^*, c^*), where the asterisk denotes complex conjugate. Thus, if $c = c_R + ic_I$ is a solution, so is $c = c_R - ic_I$. Consequently, the statement that instability occurs, i.e., there exists c for which $c_I > 0$, is equivalent to the statement that c exists for which $c_I \neq 0$.

Rayleigh and Fjortoft stability criteria

Suppose that instability occurs, i.e., $c_I > 0$. If (5.58) is multiplied by $\varphi_o^*/\left(U_s - c\right)$ and the resulting expression integrated with respect to y over the interval $(0, L)$, we obtain the balance

$$\int_0^L \left[\left|\varphi_{o_y}\right|^2 + \left(k^2 + 1\right)\left|\varphi_o\right|^2\right] dy$$

$$= \int_0^L \frac{\left(1 - U_{s_{yy}} + U_s\right)\left|\varphi_o\right|^2}{U_s - c} dy$$

$$= \int_0^L \frac{\left(1 - U_{s_{yy}} + U_s\right)\left(U_s - c^*\right)\left|\varphi_o\right|^2}{\left|U_s - c\right|^2} dy. \tag{5.60}$$

The real part of this balance is

$$\int_0^L \left[\left| \varphi_{o_y} \right|^2 + (k^2 + 1) \left| \varphi_o \right|^2 \right] dy$$

$$= \int_0^L \frac{(1 - U_{s_{yy}} + U_s)(U_s - c_R) |\varphi_o|^2}{|U_s - c|^2} dy, \qquad (5.61)$$

and the imaginary part of (5.60) can be written in the form

$$c_I \left\{ \int_0^L \frac{(1 - U_{s_{yy}} + U_s) |\varphi_o|^2}{|U_s - c|^2} dy \right\} = 0. \qquad (5.62)$$

The generalization of Rayleigh's Theorem 4.9 for the *CHM* equation is given by

Theorem 5.5 *A necessary condition for normal mode instability (i.e., $c_I > 0$) is that there exists $y^* \in (0, L)$ for which*

$$\left[1 - U_{s_{yy}} + U_s \right]_{y=y^*} = 0. \qquad (5.63)$$

Conversely, a sufficient condition for normal mode stability (i.e., $c_I = 0$) is that $1 - U_{s_{yy}} + U_s$ is of definite sign $\forall \; y \in (0, L)$.

Proof. If (5.63) does not hold for some $y \in (0, L)$, then $1 - U_{s_{yy}} + U_s$ is either positive or negative definite, that is, it is definite. But then

$$\int_0^L \frac{(1 - U_{s_{yy}} + U_s) |\varphi_o|^2}{|U_s - c|^2} dy \neq 0.$$

It therefore follows from (5.62) that $c_I = 0$, i.e., $U_s(y)$ is stable. ∎

The physics behind the stability conditions of Rayleigh's Infection point theorem and (5.63) are identical. The total vorticity, denoted here as q_{total}, associated with the two dimensional vorticity equation (3.8) is given by

$$q_{total} = \Delta \psi,$$

whereas the total vorticity associated with the *CHM* equation (5.1) is given by

$$q_{total} = \Delta \psi - \psi + y.$$

Theorems 4.9 and 5.5 are identical in that they both state, in their respective contexts, that a sufficient condition for normal mode stability of the parallel shear flow solution

$$\psi_s(y) = -\int^y U_s(\xi)\, d\xi, \tag{5.64}$$

is that the total mean flow vorticity gradient $\dfrac{\partial q^s_{total}(y)}{\partial y}$, where

$$q^s_{total}(y) \equiv [q_{total}]_{\psi=\psi_s(y)},$$

is nowhere zero in the flow domain.

Let us suppose that normal mode instability occurs. Then from (5.62), it follows that

$$\int_0^L \frac{\left(1 - U_{s_{yy}} + U_s\right)|\varphi_o|^2}{|U_s - c|^2}\, dy = 0.$$

It follows that (5.61) can be written in the form

$$\int_0^L \left[\left|\varphi_{o_y}\right|^2 + \left(k^2 + 1\right)|\varphi_o|^2 \right] dy$$

$$= \int_0^L \frac{\left(1 - U_{s_{yy}} + U_s\right) U_s\, |\varphi_o|^2}{|U_s - c|^2}\, dy.$$

Observing that the left-hand-side of this expression is positive definite for a nontrivial mode, it follows that a sufficient condition for stability for the parallel shear flow $U_s(y)$ is that

$$\left(1 - U_{s_{yy}} + U_s\right) U_s \leq 0,$$

everywhere in the flow domain. Thus, we have established the analogue of Fjortoft's Theorem 4.10 for the *CHM* equation

Theorem 5.6 *If*

$$\left(1 - U_{s_{yy}} + U_s\right) U_s \leq 0, \tag{5.65}$$

for all $y \in (0, L)$ *for the parallel shear flow* $U_s = U_s(y)$, *then* $U_s(y)$ *is a stable solution to the CHM equation with respect to normal mode perturbations.*

In terms of the total vorticity q_{total}, Fjortoft's Theorem 4.10 and Theorem 5.6 both state, in their respective contexts, that a sufficient condition for the normal mode stability of the parallel shear flow (5.64) is that the inequality

$$U_s(y) \frac{\partial q^s_{total}(y)}{\partial y} \leq 0, \tag{5.66}$$

hold everywhere in the flow domain.

Linear stability of steady solutions

The stability conditions for steady solutions of the *CHM* equation are very similar to those derived for the two dimensional vorticity equation and thus we will be brief in our presentation. Many of the proofs will be omitted.

The second variation of the constrained Hamiltonian (5.44) used in the variational principle Theorem 5.3 is given by

$$\delta^2 \mathcal{H}(q) = \iint_\Omega \left\{ [\Phi(q) - \psi] \delta^2 q + \Phi'(q)(\delta q)^2 \right.$$

$$\left. + \nabla \delta\psi \cdot \nabla \delta\psi + (\delta\psi)^2 \right\} \, dxdy,$$

where we have integrated by parts once and defined

$$\Phi'(q) \equiv \frac{d\Phi(q)}{dq}.$$

It follows that

$$\delta^2 \mathcal{H}(q_s) = \iint_\Omega \left[\Phi'_s (\delta q)^2 + \nabla \delta\psi \cdot \nabla \delta\psi + (\delta\psi)^2 \right] \, dxdy, \tag{5.67}$$

where

$$\Phi'_s \equiv \left[\frac{d\Phi(q)}{dq} \right]_{q=q_s}.$$

Theorem 5.7 *The second variation $\delta^2 \mathcal{H}(q_s)$ is an invariant of the linear stability equation (5.50), where it is understood that $\delta\psi(x, y, t)$ solves the linear stability equation.*

Proof. The proof is very similar to that given for Theorem 4.11 and thus we will be brief. We have, after an integration by parts, that

$$\frac{d\delta^2 \mathcal{H}(q_s)}{dt} = 2 \iint_\Omega [\Phi'_s \delta q - \delta\psi] \delta q_t \, dxdy.$$

If the linear stability equation (5.50) is used to replace δq_t, we have

$$\frac{d\delta^2 \mathcal{H}(q_s)}{dt} = -\iint_\Omega \partial \left(q_s, [\Phi'_s \delta q - \delta\psi]^2 \right) \, dxdy$$

$$= -\iint_\Omega \nabla \cdot \left((\mathbf{e}_3 \times \nabla q_s) \left[\Phi_s' \delta q - \delta \psi \right]^2 \right) \, dx dy$$

$$= \int_{\partial\Omega} \frac{\partial q_s}{\partial s} \left(\Phi_s' \delta q \right)^2 ds$$

$$= \int_{\partial\Omega} \frac{\partial \varphi_s}{\partial s} \Phi_s' \left(\delta q \right)^2 ds = 0. \quad \blacksquare$$

Theorem 5.8 *Let $\varphi_s(x,y)$ be a steady solution of the CHM equation (5.1) as defined by (5.43) satisfying the variational principle in Theorem 5.3. If*

$$0 < \mu_1 \le \Phi_s' \le \mu_2 < \infty, \tag{5.68}$$

for all $(x,y) \in \Omega$, where μ_1 and μ_2 are a strictly positive real numbers, then the steady solution $\varphi_s(x,y)$ is linearly stable in the sense of Liapunov with respect to the perturbation energy-enstrophy norm

$$\|\delta\varphi\|^2 = \iint_\Omega \nabla\delta\varphi \cdot \nabla\delta\varphi + (\delta\varphi)^2 + (\Delta\delta\varphi)^2 \, dx dy, \tag{5.69}$$

where $\delta\varphi(x,y,t)$ solves the linear stability equation (5.50).

Proof. Clearly, (5.68) is sufficient to ensure that $\delta^2\mathcal{H}(q_s)$ is positive definite for all nontrivial perturbations. Hence, all that remains to be established is the following estimate. From (5.69) we have

$$\|\delta\varphi\|^2 = \iint_\Omega \nabla\delta\varphi \cdot \nabla\delta\varphi + (\delta\varphi)^2 + (\Delta\delta\varphi - \delta\varphi + \delta\varphi)^2 \, dx dy$$

$$\le \iint_\Omega \nabla\delta\varphi \cdot \nabla\delta\varphi + 3(\delta\varphi)^2 + 2(\Delta\delta\varphi - \delta\varphi)^2 \, dx dy$$

$$\le 3 \iint_\Omega \nabla\delta\varphi \cdot \nabla\delta\varphi + (\delta\varphi)^2 + (\Delta\delta\varphi - \delta\varphi)^2 \, dx dy$$

$$\le \frac{3\delta^2\mathcal{H}(q_s)}{\min\{1,\mu_1\}} = \frac{3\left[\delta^2\mathcal{H}(q_s)\right]_{t=0}}{\min\{1,\mu_1\}}$$

$$\leq \frac{3\max\{1,\mu_2\}}{\min\{1,\mu_1\}} \iint\limits_{\Omega} \nabla\delta\widetilde{\varphi}\cdot\nabla\delta\widetilde{\varphi} + (\delta\widetilde{\varphi})^2 + (\triangle\delta\widetilde{\varphi} - \delta\widetilde{\varphi})^2 \, dxdy$$

$$\leq \frac{3\max\{1,\mu_2\}}{\min\{1,\mu_1\}} \iint\limits_{\Omega} \nabla\delta\widetilde{\varphi}\cdot\nabla\delta\widetilde{\varphi} + 3(\delta\widetilde{\varphi})^2 + 2(\triangle\delta\widetilde{\varphi})^2 \, dxdy$$

$$\leq \frac{9\max\{1,\mu_2\}}{\min\{1,\mu_1\}} \|\delta\widetilde{\varphi}\|^2 \, ,$$

where $\delta\widetilde{\varphi}(x,y) = \delta\varphi(x,y,0)$, and where we have used the inequality $(x+y)^2 \leq 2(x^2+y^2)$. Thus $\forall\, \varepsilon > 0$,

$$\|\delta\widetilde{\varphi}\| < \varepsilon \left[\frac{\min\{1,\mu_1\}}{9\max\{1,\mu_2\}}\right]^{\frac{1}{2}} \implies \|\delta\varphi\| < \varepsilon \,\forall\, t \geq 0. \quad\blacksquare$$

The stability condition (5.68) explicitly requires that

$$\inf_{\Omega} \left[\Phi'_s\left(q_s\left(x,y\right)\right)\right],$$

can be bounded away from zero. If this is not possible, then linear stability in the sense of Liapunov can still be proven with respect if

$$0 \leq \Phi'_s \leq \mu < \infty,$$

provided we introduce the sense of stability in Definition 4.14 understood in the context of the *CHM* equation.

Theorem 5.9 *Let $\varphi_s(x,y)$ be a steady solution of the CHM equation (5.1) as defined by (5.43) satisfying the variational principle Theorem 5.3. If*

$$0 \leq \Phi'_s \leq \mu < \infty, \tag{5.70}$$

for all $(x,y) \in \Omega$, where μ is a strictly positive real number, then the steady solution $\varphi_s(x,y)$ is linearly stable in the sense of Liapunov with respect to the perturbation energy-enstrophy norm

$$\|\delta\varphi\|_I^2 = \iint\limits_{\Omega} \nabla\delta\varphi\cdot\nabla\delta\varphi + (\delta\varphi)^2 + (\triangle\delta\varphi)^2 \, dxdy, \tag{5.71}$$

and the energy norm

$$\|\delta\varphi\|_{II}^2 = \iint\limits_{\Omega} \nabla\delta\varphi\cdot\nabla\delta\varphi + (\delta\varphi)^2 \, dxdy, \tag{5.72}$$

where $\delta\varphi(x,y,t)$ solves the linear stability equation (5.50).

Proof. Here again, (5.70) guarantees that $\delta^2 \mathcal{H}(q_s)$ positive definite for all nontrivial perturbations. We now establish the following estimate. From (5.72) we have

$$\|\delta\varphi\|_{II}^2 = \iint_\Omega \nabla\delta\varphi \cdot \nabla\delta\varphi + (\delta\varphi)^2 \; dxdy$$

$$\leq \delta^2 \mathcal{H}(q_s) = \left[\delta^2 \mathcal{H}(q_s)\right]_{t=0}$$

$$\leq \max\{1,\mu\} \iint_\Omega \nabla\delta\widetilde{\varphi} \cdot \nabla\delta\widetilde{\varphi} + (\delta\widetilde{\varphi})^2 + (\triangle\delta\widetilde{\varphi} - \delta\widetilde{\varphi})^2 \, dxdy$$

$$\leq \max\{1,\mu\} \iint_\Omega \nabla\delta\widetilde{\varphi} \cdot \nabla\delta\widetilde{\varphi} + 3\,(\delta\widetilde{\varphi})^2 + 2\,(\triangle\delta\widetilde{\varphi})^2 \, dxdy$$

$$\leq 3\max\{1,\mu\} \, \|\delta\widetilde{\varphi}\|_I^2,$$

where $\delta\widetilde{\varphi}(x,y) = \delta\varphi(x,y,0)$, and where we have used the inequality $(x+y)^2 \leq 2(x^2+y^2)$. Thus $\forall \, \varepsilon > 0$,

$$\|\delta\widetilde{\varphi}\|_I < \varepsilon \left[3\max\{1,\mu\}\right]^{-\frac{1}{2}} \implies \|\delta\varphi\|_{II} < \varepsilon \, \forall \, t \geq 0. \; \blacksquare$$

The analogue of Theorem 4.17 for the *CHM* equation requires bounding the kinetic *and* potential energy functional by the enstrophy functional. It is possible to do this by first establishing the following Poincaré Inequality between the potential and kinetic energy functionals.

Lemma 5.10 *Suppose that the minimum eigenvalue, denoted $\lambda_{\min}$, of the boundary-value problem*

$$\triangle\phi + \lambda\phi = 0, \; (x,y) \in \Omega,$$

$$\phi|_{\partial\Omega} = 0,$$

is strictly positive. If $h(x,y)$ is any twice-continuously differentiable function satisfying $h|_{\partial\Omega} = 0$, then

$$\iint_\Omega h^2 \, dxdy \leq \frac{1}{\lambda_{\min}} \iint_\Omega \nabla h \cdot \nabla h \, dxdy.$$

Proof. We use a *Rayleigh-Ritz* variational argument similar to that used for Lemma 4.16. If

$$\triangle\phi + \lambda\phi = 0,$$

is multiplied through by ϕ and the first term integrated by parts, it follows that

$$\lambda = \frac{\displaystyle\iint_\Omega \nabla\phi \cdot \nabla\phi \, dxdy}{\displaystyle\iint_\Omega \phi^2 \, dxdy},$$

where we have used $\phi|_{\partial\Omega} = 0$. Thus a *Rayleigh-Ritz* variational principle for $\lambda_{\min}$ is

$$\lambda_{\min} = \min_\phi \left\{ \frac{\displaystyle\iint_\Omega \nabla\phi \cdot \nabla\phi \, dxdy}{\displaystyle\iint_\Omega \phi^2 \, dxdy} \right\},$$

where the minimization is done over all twice-continuously differentiable nontrivial functions $\phi(x,y)$ which satisfy the boundary condition $\phi|_{\partial\Omega} = 0$. It follows that if $h(x,y)$ is an arbitrary twice-continuously differentiable function satisfying $h|_{\partial\Omega} = 0$, then

$$\frac{\displaystyle\iint_\Omega \nabla h \cdot \nabla h \, dxdy}{\displaystyle\iint_\Omega h^2 \, dxdy} \geq \lambda_{\min},$$

which can be re-arranged into

$$\iint_\Omega h^2 \, dxdy \leq \frac{1}{\lambda_{\min}} \iint_\Omega \nabla h \cdot \nabla h \, dxdy. \quad \blacksquare$$

Lemma 5.11 *Suppose that the minimum eigenvalue, denoted $\lambda_{\min}$, of the boundary-value problem*

$$\triangle\phi + \lambda\phi = 0, \quad (x,y) \in \Omega,$$

$$\phi|_{\partial\Omega} = 0,$$

is strictly positive. If $h(x,y)$ is any twice-continuously differentiable function satisfying $h|_{\partial\Omega} = 0$, then

$$\iint_\Omega h^2 \, dxdy \leq \frac{1}{\lambda_{\min}^2} \iint_\Omega (\triangle h)^2 \, dxdy.$$

Proof. The result follows straightforwardly from Lemmas 5.10 and 4.16. ∎

Theorem 5.12 *Suppose that the domain Ω is bounded in at least one direction. Let $\varphi_s(x,y)$ be a steady solution of the CHM equation (5.1) as defined by (5.43) satisfying the variational principle in Theorem 5.3. If*

$$-\infty < \mu_1 \leq \Phi'_s \leq \mu_2 < -\frac{\lambda_{\min}+1}{\lambda_{\min}^2} < 0, \tag{5.73}$$

for all $(x,y) \in \Omega$, where μ_1 and μ_2 are strictly negative real numbers where $\lambda_{\min}$ is the strictly positive minimum eigenvalue of the boundary-value problem

$$\Delta\phi + \lambda\phi = 0, \quad (x,y) \in \Omega,$$

$$\phi|_{\partial\Omega} = 0,$$

then the steady solution $\varphi_s(x,y)$ is linearly stable in the sense of Liapunov with respect to the perturbation enstrophy norm

$$\|\delta\varphi\|^2 = \iint_{\Omega} (\Delta\delta\varphi)^2 \, dxdy, \tag{5.74}$$

where $\delta\varphi(x,y,t)$ solves the linear stability equation (5.50).

Proof. First, we shall bound the perturbation norm in terms of $\delta^2\mathcal{H}(q_s)$. We have

$$\delta^2\mathcal{H}(q_s) = \iint_{\Omega} \nabla\delta\varphi \cdot \nabla\delta\varphi + (\delta\varphi)^2 + \Phi'_s (\Delta\delta\varphi - \delta\varphi)^2 \, dxdy$$

$$\leq \iint_{\Omega} \left(\frac{\lambda_{\min}+1}{\lambda_{\min}^2}\right) (\Delta\delta\varphi)^2 + \Phi'_s (\Delta\delta\varphi - \delta\varphi)^2 \, dxdy$$

$$\leq \iint_{\Omega} \left(\frac{\lambda_{\min}+1}{\lambda_{\min}^2}\right) (\Delta\delta\varphi)^2 + \mu_2 (\Delta\delta\varphi - \delta\varphi)^2 \, dxdy$$

$$= \iint_{\Omega} \left(\frac{\lambda_{\min}+1}{\lambda_{\min}^2} + \mu_2\right) (\Delta\delta\varphi)^2 + \mu_2 \left[2\nabla\delta\varphi \cdot \nabla\delta\varphi + (\delta\varphi)^2\right] \, dxdy$$

$$\leq \Gamma_1 \|\delta\varphi\|^2,$$

where we assume

$$\Gamma_1 \equiv \frac{\lambda_{\min} + 1}{\lambda_{\min}^2} + \mu_2 < 0,$$

where we integrated by parts once assuming $\delta\varphi|_{\partial\Omega} = 0$. This inequality may be re-arranged into the form

$$\|\delta\varphi\|^2 \leq \Gamma_1^{-1}\delta^2\mathcal{H}(q_s) = \Gamma_1^{-1}\left[\delta^2\mathcal{H}(q_s)\right]_{t=0}$$

$$= \Gamma_1^{-1} \iint_{\Omega} \nabla\delta\widetilde{\varphi}\cdot\nabla\delta\widetilde{\varphi} + (\delta\widetilde{\varphi})^2 + \Phi_s'(\Delta\delta\widetilde{\varphi} - \delta\widetilde{\varphi})^2 \, dxdy$$

$$\leq \Gamma_1^{-1} \iint_{\Omega} \Phi_s'(\Delta\delta\widetilde{\varphi} - \delta\widetilde{\varphi})^2 \, dxdy$$

$$\leq \frac{\mu_1}{\Gamma_1} \iint_{\Omega} (\Delta\delta\widetilde{\varphi} - \delta\widetilde{\varphi})^2 \, dxdy$$

$$= \frac{\mu_1}{\Gamma_1} \iint_{\Omega} \left[(\Delta\delta\widetilde{\varphi})^2 + 2\nabla\delta\widetilde{\varphi}\cdot\nabla\delta\widetilde{\varphi} + (\delta\widetilde{\varphi})^2\right] \, dxdy$$

$$\leq \Gamma_2 \|\delta\widetilde{\varphi}\|^2,$$

where

$$\Gamma_2 \equiv \frac{\mu_1\left(1 + 2\lambda_{\min} + \lambda_{\min}^2\right)}{\Gamma_1\lambda_{\min}^2} > 0,$$

and where $\widetilde{\varphi}(x, y) = \varphi(x, y, 0)$. Therefore, for every $\varepsilon > 0$, $\|\delta\widetilde{\varphi}\| < \varepsilon\Gamma_2^{-\frac{1}{2}} \Longrightarrow \|\delta\varphi\| < \varepsilon$ for all $t \geq 0$. $\blacksquare$

Reduction to Fjortoft's and Rayleigh's stability theorems

Theorem 5.9 reduces to Fjortoft's Stability Theorem 5.6 for a parallel shear flow $\varphi_s = \varphi_s(y)$. If (5.43) is differentiated with respect to y we obtain

$$\Phi_s' = \frac{U_s}{U_{s_{yy}} - U_s - 1}. \tag{5.75}$$

Thus, for the parallel shear flow $\varphi_s = \varphi_s(y)$, it follows that

$$\Phi'_s = \frac{U_s}{U_{s_{yy}} - U_s - 1} \geq 0$$

$$\Longrightarrow U_s \left(U_{s_{yy}} - U_s - 1 \right) \geq 0$$

$$\Longleftrightarrow U_s \left(1 - U_{s_{yy}} + U_s \right) \leq 0,$$

which is exactly the stability condition stated in Theorem 5.6.

Parallel shear flows of the form $\psi = \varphi_s(y)$ in the periodic channel domain given by (5.46) and (5.47) satisfy the variational principle given in Theorem 5.4. This variational principle can be exploited to derive Rayleigh's Stability Theorem 5.5.

Lemma 5.13 *The second variation $\delta^2 \mathcal{L}(q_s)$, where $\mathcal{L}(q)$ is the functional associated with the variational principle Theorem 5.4, is an invariant of the linear stability equation (5.50).*

Proof. The proof is identical to that of Lemma 4.27 and is left as an exercise for the reader. ∎

Theorem 5.14 *The parallel shear solution $\psi = \varphi_s(y)$ (in the periodic channel domain (5.46) satisfying the boundary conditions (5.47) and the smooth periodicity conditions at $x = \pm x_o$) is linearly stable in the sense of Liapunov with respect to the perturbation enstrophy norm*

$$\|\delta\varphi\|^2 = \iint\limits_\Omega (\delta q)^2 \ dxdy,$$

if

$$\mathcal{Y}'(q_s) \equiv \left[\frac{d\mathcal{Y}(q)}{dq} \right]_{q=q_s}$$

is of definite sign, where $\mathcal{Y}(q)$ is the inverse function associated with

$$q = \varphi_{s_{yy}}(y) - \varphi_s(y) + y,$$

and where $\delta\varphi(x, y, t)$ solves the linear stability equation (5.50) and $\delta q = (\triangle - 1)\delta\varphi$.

Proof. The proof is identical to that of Theorem 4.28 and is left as an exercise for the reader. ■

It is straightforward to show that the definiteness of $\mathcal{Y}'(q_s)$ is equivalent to Rayleigh's Stability Theorem 5.5. It follows from the definition of $\mathcal{Y}(q)$ that

$$\mathcal{Y}\left(\varphi_{s_{yy}}(y) - \varphi_s(y) + y\right) = y.$$

If this expression is differentiated with respect to y we obtain

$$\mathcal{Y}'(q_s) = \frac{1}{1 - U_{s_{yy}} + U_s}, \quad (q_{s_y} \neq 0).$$

It therefore follows that if $\mathcal{Y}'(q_s)$ is definite so is $1 - U_{s_{yy}} + U_s$.

Andrews's Theorem

As in the stability theory developed for the two dimensional vorticity equation, there may be symmetry restrictions on the class of flows that can, in principle, satisfy the stability conditions (5.70) or (5.73). For example, we now show that the only steady flows which satisfy (5.70) or (5.73) in a periodic channel domain are parallel shear flows. Here, again, we emphasize that this does *not* imply that flows which do not satisfy the stability conditions (5.70) or (5.73) are necessarily unstable.

Theorem 5.15 *The only steady solutions for the periodic channel domain Ω given by (5.46) satisfying the boundary conditions (5.47) and the smooth periodicity conditions at $x = \pm x_o$ which can satisfy the linear stability criteria (5.70) or (5.73) are the parallel shear flows $\varphi_s = \varphi_s(y)$.*

Proof. Let $\varphi_s(x, y)$ be an arbitrary steady solution to the *CHM* equation (5.1) given by

$$\varphi_s = \Phi\left(\Delta\varphi_s - \varphi_s + y\right),$$

it follows that

$$\frac{\partial\varphi_s}{\partial x} = (\Delta\varphi_s - \varphi_s)_x \, \Phi'_s.$$

If this expression is multiplied through by $(\Delta\varphi_s - \varphi_s)_x$ and the result integrated over Ω we obtain

$$\iint_\Omega \nabla\left(\frac{\partial\varphi_s}{\partial x}\right) \cdot \nabla\left(\frac{\partial\varphi_s}{\partial x}\right) + \left(\frac{\partial\varphi_s}{\partial x}\right)^2 + [(\Delta\varphi_s - \varphi_s)_x]^2 \, \Phi'_s \, dx dy$$

$$= \int_{-x_o}^{x_o} \left[\frac{\partial \varphi_s}{\partial x}\frac{\partial^2 \varphi_s}{\partial x \partial y}\right]_{y=L} - \left[\frac{\partial \varphi_s}{\partial x}\frac{\partial^2 \varphi_s}{\partial x \partial y}\right]_{y=0} dx$$

$$+ \int_0^L \left[\frac{\partial \varphi_s}{\partial x}\frac{\partial^2 \varphi_s}{\partial x^2}\right]_{x=x_o} - \left[\frac{\partial \varphi_s}{\partial x}\frac{\partial^2 \varphi_s}{\partial x^2}\right]_{x=-x_o} dy, \tag{5.76}$$

where we have integrated by parts. However, exploiting the periodicity of φ_s and

$$\left[\frac{\partial \varphi_s}{\partial x}\right]_{y=0} = \left[\frac{\partial \varphi_s}{\partial x}\right]_{y=L} = 0,$$

it follows from (5.76) that

$$\iint_\Omega \left\{\nabla\left(\frac{\partial \varphi_s}{\partial x}\right) \cdot \nabla\left(\frac{\partial \varphi_s}{\partial x}\right) + \left(\frac{\partial \varphi_s}{\partial x}\right)^2\right.$$

$$\left. + \left[(\Delta\varphi_s - \varphi_s)_x\right]^2 \Phi_s' \right\} dxdy = 0. \tag{5.77}$$

However, if

$$0 \le \Phi_s' < \mu < \infty, \ \forall \ (x,y) \in \Omega,$$

it follows from (5.77) that

$$\frac{\partial \varphi_s}{\partial x} \equiv 0 \implies \varphi_s = \varphi_s(y).$$

Thus we have shown that the only steady flows in the periodic channel that can satisfy (5.70) are parallel shear flows of the form $\varphi_s = \varphi_s(y)$.

Assuming that φ_s is sufficiently smooth, it follows that $\dfrac{\partial \varphi_s}{\partial x}$ satisfies the Poincaré Inequalities (Lemmas 4.15 and 5.10 applied to the periodic channel)

$$\iint_\Omega \nabla\varphi_{s_x} \cdot \nabla\varphi_{s_x} \, dxdy \le \left(\frac{L}{\pi}\right)^2 \iint_\Omega (\Delta\varphi_{s_x})^2 \, dxdy,$$

$$\iint_\Omega (\varphi_{s_x})^2 \, dxdy \le \left(\frac{L}{\pi}\right)^4 \iint_\Omega (\Delta\varphi_{s_x})^2 \, dxdy.$$

Substituting these inequalities into (5.77) implies

$$0 \le \iint_\Omega \left[\left(\frac{L}{\pi} \right)^2 + \left(\frac{L}{\pi} \right)^4 \right] (\Delta \varphi_{s_x})^2 + (\Delta \varphi_{s_x} - \varphi_{s_x})^2 \, \Phi_s' \, dx dy.$$

But if (5.73) holds, i.e.,

$$-\infty < \mu_1 \le \Phi_s' \le \mu_2 < - \left[\left(\frac{L}{\pi} \right)^2 + \left(\frac{L}{\pi} \right)^4 \right] < 0,$$

$\forall \, (x, y) \in \Omega$, then this inequality implies

$$0 \le \iint_\Omega \left[\left(\frac{L}{\pi} \right)^2 + \left(\frac{L}{\pi} \right)^4 \right] (\Delta \varphi_{s_x})^2 + (\Delta \varphi_{s_x} - \varphi_{s_x})^2 \, \Phi_s' \, dx dy$$

$$\le \iint_\Omega \left\{ \left[\left(\frac{L}{\pi} \right)^2 + \left(\frac{L}{\pi} \right)^4 + \mu_2 \right] (\Delta \varphi_{s_x})^2 \right.$$

$$\left. + \mu_2 \left[2 \nabla \varphi_{s_x} \cdot \nabla \varphi_{s_x} + (\varphi_{s_x})^2 \right] \right\} \, dx dy$$

$$\le \iint_\Omega \left[\left(\frac{L}{\pi} \right)^2 + \left(\frac{L}{\pi} \right)^4 + \mu_2 \right] (\Delta \varphi_{s_x})^2 \, dx dy \le 0.$$

Thus, we have shown that if (5.73) holds, then

$$0 \le \iint_\Omega (\Delta \varphi_{s_x})^2 \, dx dy \le 0.$$

However, this inequality implies that

$$\frac{\partial \Delta \varphi_s}{\partial x} = 0,$$

$\forall \, (x, y) \in \Omega$. If this expression is multiplied through by $\dfrac{\partial \varphi_s}{\partial x}$ and the result integrated over Ω we obtain

$$\iint_\Omega \nabla \left(\frac{\partial \varphi_s}{\partial x} \right) \cdot \nabla \left(\frac{\partial \varphi_s}{\partial x} \right) \, dx dy = 0,$$

where we have exploited the periodicity of φ_s at $x = \pm x_o$ and the boundary conditions on $y = 0$ and $y = L$. It follows from this expression that

$$\frac{\partial \varphi_s}{\partial x} \equiv 0 \Longrightarrow \varphi_s = \varphi_s(y).$$

Thus, the only steady flows which can satisfy (5.73) in the periodic channel domain are the parallel shear flows $\varphi_s = \varphi_s(y)$. ∎

Nonlinear stability of steady solutions

The nonlinear stability theory for arbitrary steady solutions to the *CHM* equation requires introducing the functional $\mathcal{N}(q)$ given by

$$\mathcal{N}(q) = \mathcal{H}(q_s + q) - \mathcal{H}(q_s) - \delta\mathcal{H}(q_s), \tag{5.78}$$

where $\mathcal{H}(q)$ is the functional associated with the variational principle Theorem 5.3 and where $\delta\mathcal{H}(q_s)$ is written in the form

$$\delta\mathcal{H}(q_s) = \iint_{\Omega} (\Phi(q_s) - \psi_s) q \, dx dy = 0. \tag{5.79}$$

The *total* vorticity, given by

$$q_T(x, y, t) \equiv q_s(x, y) + q(x, y, t),$$

solves the full nonlinear *CHM* equation (5.1). We will call $q(x, y, t)$ the disturbance or perturbation vorticity.

Clearly, the functional $\mathcal{N}(q)$ is an invariant of the *CHM* equation since $\mathcal{H}(q_T)$ is an invariant, $\mathcal{H}(q_s)$ is time independent and $\delta\mathcal{H}(q_s) = 0$. Consequently, it follows that

$$\mathcal{N}(q) = \mathcal{N}(\tilde{q}), \tag{5.80}$$

where $\tilde{q}(x, y) \equiv q(x, y, 0)$. In addition, observe that $\mathcal{N}(0) = 0$. If $\mathcal{N}(q)$ is Taylor expanded about $q = 0$, it follows that

$$\mathcal{N}(q) = \frac{1}{2}\delta^2\mathcal{H}(q_s) + h.o.t., \tag{5.81}$$

where the variations in $\delta^2\mathcal{H}(q_s)$ are replaced with the disturbance vorticity $q(x, y, t)$.

Substituting the constrained Hamiltonian $\mathcal{H}$ given by (5.44) into (5.78), implies

$$\mathcal{N}(q) = \iint_{\Omega} \frac{\nabla\varphi \cdot \nabla\varphi + \varphi^2}{2} + \int_{q_s}^{q_s+q} \Phi(\xi) \, d\xi - \Phi(q_s) q \, dx dy, \tag{5.82}$$

where

$$q_s \equiv \Delta\varphi_s - \varphi_s + y \text{ and } q \equiv \Delta\varphi - \varphi,$$

and we have integrated by parts once.

Lemma 5.16 *Suppose there exist real numbers α and β such that*

$$\alpha \leq \Phi'(q) \leq \beta,$$

for all q, then

$$\frac{\alpha q^2}{2} \leq \int_{q_s}^{q_s+q} \Phi\left(\xi\right) d\xi - \Phi\left(q_s\right) q \;\leq\; \frac{\beta q^2}{2}, \tag{5.83}$$

for all q_s and q.

Proof. The Lemma is simply an application of Lemma 4.19. The details are left as an exercise for the reader. ∎

Theorem 5.17 *Let $\varphi_s(x,y)$ be a steady solution of the CHM equation (5.1) as defined by (5.43) satisfying the variational principle Theorem 5.3. If*

$$0 < \mu_1 \leq \Phi'(q) \leq \mu_2 < \infty, \tag{5.84}$$

for all q, where μ_1 and μ_2 are a strictly positive real numbers, then the steady solution $\varphi_s(x,y)$ is nonlinearly stable in the sense of Liapunov with respect to the disturbance energy-enstrophy norm

$$\|\varphi\|^2 = \iint_\Omega \nabla\varphi \cdot \nabla\varphi + \varphi^2 + (\Delta\varphi)^2 \, dxdy. \tag{5.85}$$

Proof. It follows from (5.85) that

$$\|\varphi\|^2 = \iint_\Omega \nabla\varphi \cdot \nabla\varphi + \varphi^2 + (\Delta\varphi)^2 \, dxdy$$

$$\leq 3 \iint_\Omega \nabla\varphi \cdot \nabla\varphi + \varphi^2 + (\Delta\varphi - \varphi)^2 \, dxdy$$

$$\leq \frac{3}{\min\left[1,\mu_1\right]} \iint_\Omega \nabla\varphi \cdot \nabla\varphi + \varphi^2 + \mu_1 (\Delta\varphi - \varphi)^2 \, dxdy$$

$$\leq \frac{6\mathcal{N}(q)}{\min\left[1,\mu_1\right]} = \frac{6\mathcal{N}(\tilde{q})}{\min\left[1,\mu_1\right]}$$

$$\leq \frac{3}{\min\left[1,\mu_1\right]} \iint_\Omega \nabla\widetilde{\varphi}\cdot\nabla\widetilde{\varphi}+\widetilde{\varphi}^2+\mu_2\left(\Delta\widetilde{\varphi}-\widetilde{\varphi}\right)^2 dxdy$$

$$\leq \frac{3\max\left[1,\mu_2\right]}{\min\left[1,\mu_1\right]} \iint_\Omega \nabla\widetilde{\varphi}\cdot\nabla\widetilde{\varphi}+\widetilde{\varphi}^2+\left(\Delta\widetilde{\varphi}-\widetilde{\varphi}\right)^2 dxdy$$

$$\leq \frac{9\max\left[1,\mu_2\right]}{\min\left[1,\mu_1\right]} \iint_\Omega \nabla\widetilde{\varphi}\cdot\nabla\widetilde{\varphi}+\widetilde{\varphi}^2+\left(\Delta\widetilde{\varphi}\right)^2 dxdy$$

$$= \frac{9\max\left[1,\mu_2\right]}{\min\left[1,\mu_1\right]} \|\widetilde{\varphi}\|^2\,,$$

where we have used (5.83), the inequality $(x+y)^2 \leq 2\left(x^2+y^2\right)$, and defined $\widetilde{\varphi}\left(x,y\right) = \varphi\left(x,y,0\right)$. Thus, we conclude that for every $\varepsilon > 0$

$$\|\widetilde{\varphi}\| < \varepsilon \left[\frac{9\max\left(1,\mu_2\right)}{\min\left(1,\mu_1\right)}\right]^{-\frac{1}{2}} \implies \|\varphi\| < \varepsilon \;\forall\, t \geq 0. \;\blacksquare$$

The stability conditions (5.84) explicitly require that μ_1 is strictly greater than zero. If this is not possible, then nonlinear stability in the sense of Liapunov can still be proven if

$$0 \leq \Phi'(q) \leq \mu < \infty,$$

for all q provided we introduce the notion of stability in Definition 4.21 understood in the context of the nonlinear *CHM* equation.

Theorem 5.18 *Let $\varphi_s(x,y)$ be a steady solution of the CHM equation (5.1) as defined by $\varphi_s = \Phi\left(\Delta\varphi_s - \varphi_s + y\right)$ satisfying the variational principle in Theorem 5.3. If*

$$0 \leq \Phi'(q) \leq \mu < \infty, \tag{5.86}$$

for all q, where μ is a strictly positive real number, then the steady solution $\varphi_s(x,y)$ is nonlinearly stable in the sense of Liapunov with respect to the disturbance energy-enstrophy norm

$$\|\varphi\|_I^2 = \iint_\Omega \nabla\varphi\cdot\nabla\varphi+\varphi^2+\left(\Delta\varphi\right)^2 dxdy, \tag{5.87}$$

and the disturbance energy norm

$$\|\varphi\|_{II}^2 = \iint_\Omega \nabla\varphi\cdot\nabla\varphi+\varphi^2 dxdy. \tag{5.88}$$

Proof. It follows from (5.86) and Lemma 5.16 that

$$0 \leq \int_{q_s}^{q_s+q} \Phi\left(\xi\right) d\xi - \Phi\left(q_s\right) q \leq \frac{\mu q^2}{2},$$

for all q. It therefore follows that

$$\|\varphi\|_{II}^2 = \iint_\Omega \nabla\varphi \cdot \nabla\varphi + \varphi^2 \, dxdy$$

$$\leq 2\mathcal{N}\left(q\right) = 2\mathcal{N}\left(\widetilde{q}\right)$$

$$\leq \iint_\Omega \nabla\widetilde{\varphi} \cdot \nabla\widetilde{\varphi} + \widetilde{\varphi}^2 + \mu\left(\Delta\widetilde{\varphi} - \widetilde{\varphi}\right)^2 dxdy$$

$$\leq 3 \max\left[1,\mu\right] \iint_\Omega \nabla\widetilde{\varphi} \cdot \nabla\widetilde{\varphi} + \widetilde{\varphi}^2 + \left(\Delta\widetilde{\varphi}\right)^2 dxdy$$

$$= 3 \max\left[1,\mu\right] \|\widetilde{\varphi}\|_I^2,$$

where $\widetilde{\varphi}(x,y) = \varphi\left(x,y,0\right)$ and we have used the inequality $\left(x+y\right)^2 \leq 2\left(x^2+y^2\right)$. Thus we conclude that for every $\varepsilon > 0$,

$$\|\widetilde{\varphi}\|_I < \varepsilon \left(3 \max\left[1,\mu\right]\right)^{-\frac{1}{2}} \implies \|\varphi\|_{II} < \varepsilon \; \forall \, t \geq 0. \quad \blacksquare$$

We complete our discussion on the stability of arbitrary solutions to the *CHM* equation by giving the nonlinear generalization of Theorem 5.12. We remind the reader that we impose the *natural* finite-amplitude disturbance boundary condition

$$\varphi = 0 \text{ on } \partial\Omega. \tag{5.89}$$

Theorem 5.19 *Suppose that the domain Ω is bounded in at least one direction. Let $\varphi_s(x,y)$ be a steady solution of the CHM equation (5.1) as defined by (5.43) satisfying the variational principle Theorem 5.3. If*

$$-\infty < \mu_1 \leq \Phi'\left(q\right) \leq \mu_2 < -\frac{\lambda_{\min}+1}{\lambda_{\min}^2} < 0, \tag{5.90}$$

for all q, where μ_1 and μ_2 are strictly negative real numbers where $\lambda_{\min}$ is the strictly positive minimum eigenvalue of the boundary-value problem

$$\left. \begin{array}{r} \Delta\phi + \lambda\phi = 0, \ (x,y) \in \Omega, \\ \phi|_{\partial\Omega} = 0, \end{array} \right\}, \tag{5.91}$$

then the steady solution $\varphi_s(x,y)$ is nonlinearly stable in the sense of Liapunov with respect to the disturbance enstrophy norm

$$\|\varphi\|^2 = \iint\limits_{\Omega} (\Delta\varphi)^2 \, dxdy. \tag{5.92}$$

Proof. First, we note that (5.90) and Lemma 5.16 implies that

$$\frac{\mu_1 q^2}{2} \leq \int\limits_{q_s}^{q_s+q} \Phi(\xi) \, d\xi - \Phi(q_s) \, q \leq \frac{\mu_2 q^2}{2},$$

for all q. We can bound the perturbation norm in terms of $\mathcal{N}(q)$ as follows. From (5.82)

$$\mathcal{N}(q) = \iint\limits_{\Omega} \frac{\nabla\varphi \cdot \nabla\varphi + \varphi^2}{2} + \int\limits_{q_s}^{q_s+q} \Phi(\xi) \, d\xi - \Phi(q_s) \, q \; dxdy$$

$$\leq \iint\limits_{\Omega} \left(\frac{\lambda_{\min}+1}{2\lambda_{\min}^2} \right) (\Delta\varphi)^2 + \int\limits_{q_s}^{q_s+q} \Phi(\xi) \, d\xi - \Phi(q_s) \, q \; dxdy$$

$$\leq \frac{1}{2} \iint\limits_{\Omega} \left(\frac{\lambda_{\min}+1}{\lambda_{\min}^2} \right) (\Delta\varphi)^2 + \mu_2 (\Delta\varphi - \varphi)^2 \, dxdy$$

$$= \frac{1}{2} \iint\limits_{\Omega} \left(\frac{\lambda_{\min}+1}{\lambda_{\min}^2} + \mu_2 \right) (\Delta\varphi)^2 + \mu_2 \left[2\nabla\varphi \cdot \nabla\varphi + \varphi^2 \right] \, dxdy$$

$$\leq \frac{\Gamma_1}{2} \|\varphi\|^2,$$

where we assume (because of (5.90))

$$\Gamma_1 \equiv \frac{\lambda_{\min}+1}{\lambda_{\min}^2} + \mu_2 < 0.$$

This inequality may be re-arranged into the form

$$\|\varphi\|^2 \leq 2\Gamma_1^{-1}\mathcal{N}(q) = 2\Gamma_1^{-1}\mathcal{N}(\tilde{q})$$

$$= 2\Gamma_1^{-1} \iint_\Omega \frac{\nabla\tilde{\varphi}\cdot\nabla\tilde{\varphi} + \tilde{\varphi}^2}{2} + \int_{q_s}^{q_s+\tilde{q}} \Phi(\xi)\,d\xi - \Phi(q_s)\,\tilde{q}\,dxdy$$

$$\leq \Gamma_1^{-1} \iint_\Omega \nabla\tilde{\varphi}\cdot\nabla\tilde{\varphi} + \tilde{\varphi}^2 + \mu_1\left(\Delta\tilde{\varphi} - \tilde{\varphi}\right)^2\,dxdy$$

$$\leq \frac{\mu_1}{\Gamma_1} \iint_\Omega \left[(\Delta\tilde{\varphi})^2 + 2\nabla\tilde{\varphi}\cdot\nabla\tilde{\varphi} + \tilde{\varphi}^2\right]\,dxdy$$

$$\leq \Gamma_2\|\delta\tilde{\varphi}\|^2,$$

where

$$\Gamma_2 \equiv \frac{\mu_1\left(1 + 2\lambda_{\min} + \lambda_{\min}^2\right)}{\Gamma_1\lambda_{\min}^2} > 0,$$

and where $\tilde{\varphi}(x,y) = \varphi(x,y,0)$. Therefore, for every $\varepsilon > 0$, $\|\delta\tilde{\varphi}\| < \varepsilon\Gamma_2^{-\frac{1}{2}} \implies \|\delta\varphi\| < \varepsilon$ for all $t \geq 0$. ∎

The nonlinear generalization of Rayleigh's inflection point theorem (Theorem 5.14) for a parallel shear flow is given as follows. First, let us introduce the functional $\mathcal{M}(q)$ given by

$$\mathcal{M}(q) = \mathcal{L}(q_s + q) - \mathcal{L}(q_s) - \delta\mathcal{L}(q_s), \qquad (5.93)$$

where $\mathcal{L}(q)$ is given by (5.48) and where $\delta\mathcal{L}(q_s)$ is written in the form

$$\delta\mathcal{L}(q_s) = \iint_\Omega \{\mathcal{Y}(q_s) - y\}\,q\,dxdy. \qquad (5.94)$$

The *total* stream function, given by

$$\psi(x,y,t) = \varphi_s(y) + \varphi(x,y,t),$$

solves the nonlinear *CHM* equation (5.1) and it is assumed that the steady solution is described by $\mathcal{Y}(q_s(y)) = y$ so that $\delta\mathcal{L}(q_s) = 0$. It is also assumed that the domain is given by the periodic channel (5.46) with the boundary conditions (5.47) where it is assumed that the total stream function is smoothly periodic at $x = \pm x_o$.

The disturbance stream function is given by $\varphi(x, y, t)$ and is assumed to satisfy the boundary conditions $\varphi(x, 0, t) = \varphi(x, L, t) = 0$ and is assumed to be smoothly periodic at $x = \pm x_o$. Since $\mathcal{M}(q)$ is the sum of three individually invariant functionals, it follows $\mathcal{M}(\varphi)$ is itself an invariant of the nonlinear *CHM* equation. Hence

$$\mathcal{M}(q(x, y, t)) = \mathcal{M}(\widetilde{q}(x, y)),$$

where $\widetilde{q}(x, y) = q(x, y, 0)$ and $\widetilde{\varphi}(x, y) = \varphi(x, y, 0)$.

Substitution of (5.48) and (5.94) into (5.93) yields

$$\mathcal{M}(q) = \iint_{\Omega} \left\{ \int_q^{q_s+q} \mathcal{Y}(\xi) \, d\xi - \mathcal{Y}(q_s) q \right\} \, dx dy. \tag{5.95}$$

The nonlinear generalization of Theorem 5.12 will require assuming that there exist finite real numbers α and β such that

$$\alpha \le \frac{d\mathcal{Y}(q)}{dq} \le \beta, \tag{5.96}$$

for all $q \in \mathbb{R}$. If (5.96) holds, then it follows from Lemma 5.16 that

$$\frac{\alpha q^2}{2} \le \int_{q_s}^{q_s+q} \mathcal{Y}(\xi) \, d\xi - \mathcal{Y}(q_s) q \le \frac{\beta q^2}{2}, \tag{5.97}$$

for all disturbance vorticity $q = (\Delta - 1)\varphi$.

Theorem 5.20 *If*

$$0 < \alpha \le \frac{d\mathcal{Y}(q)}{dq} \le \beta < \infty, \tag{5.98}$$

or if

$$-\infty < \alpha \le \frac{d\mathcal{Y}(q)}{dq} \le \beta < 0, \tag{5.99}$$

for all $q \in \mathbb{R}$, then the parallel shear flow solution $\psi = \varphi_s(y)$ is nonlinearly stable in the sense of Liapunov with respect to the disturbance enstrophy norm

$$\|\varphi\|^2 = \iint_{\Omega} (\Delta\varphi - \omega)^2 \, dx dy. \tag{5.100}$$

Proof. It follows from (5.100), (5.98), (5.97) and the invariance of $\mathcal{M}(q)$ that

$$\frac{\alpha}{2}\|\varphi\|^2 = \iint_\Omega \frac{\alpha(\Delta\varphi - \varphi)^2}{2}\, dx dy$$

$$\leq \mathcal{M}(q) = \mathcal{M}(\tilde{q}) \leq \frac{\beta}{2}\|\tilde{\varphi}\|^2 < \infty.$$

Thus for every $\varepsilon > 0$,

$$\|\tilde{\varphi}\| \leq \varepsilon(\alpha/\beta)^{\frac{1}{2}} \implies \|\varphi\| \leq \varepsilon \,\forall\, t \geq 0,$$

if (5.98) holds.

On the other hand if (5.99) holds, then

$$-\infty < \alpha\|\tilde{\varphi}\|^2 \leq 2\mathcal{N}(\tilde{q})$$

$$= 2\mathcal{N}(q) \leq \beta\|\varphi\|^2 < 0.$$

Thus, for every $\varepsilon > 0$,

$$\|\tilde{\varphi}\| \leq \varepsilon(\beta/\alpha)^{\frac{1}{2}} \implies \|\varphi\| \leq \varepsilon \,\forall\, t \geq 0,$$

if (5.99) holds. ■

5.5 Steadily-travelling solutions

In addition to steady solutions, the *CHM* equation also has steadily-travelling solutions. By a steadily-travelling solution we mean a solution to (5.1) of the form

$$\psi(x, y, t) = \varphi_s(x - ct, y). \tag{5.101}$$

If a steadily-travelling solution of the form (5.101) is substituted into (5.1), it follows that

$$\partial(\varphi_s + cy,\ \Delta\varphi_s - \varphi_s + y) = 0, \tag{5.102}$$

where it is understood that $\partial(A, B) \equiv A_\xi B_y - A_y B_\xi$ and $\Delta \equiv \partial_{\xi\xi} + \partial_{yy}$ where $\xi = x - ct$ so that we may write $\varphi_s = \varphi_s(\xi, y)$. We can immediately integrate (5.102) to yield

$$\varphi_s + cy = \Phi(\Delta\varphi_s - \varphi_s + y), \tag{5.103}$$

where $\Phi(*)$ is an arbitrary function of its argument. This expression states simply that steadily-travelling solutions possess the property that the *streak function*, given by $\varphi_s + cy$, can be written as a function of the total or potential vorticity $\Delta\varphi_s - \varphi_s + y$ in the co-moving frame of reference and represents a generalization of (4.2). We will use (5.103) to derive a variational principle for these steadily-travelling solutions.

It is important to note that the form (5.101) explicitly excludes solutions which have a component of propagation in the y-direction. We show that such solutions cannot exist if there are closed streak lines.. Streak lines are contours upon which the streak function is constant. Streak lines are the generalization of the streamline for the steadily-travelling solution. Closed streak line solutions are very important physically since they imply that fluid is being transported in the direction of propagation. As such, this class of solutions provides models for isolated steadily-travelling eddies (we will present such a solution momentarily). The argument presented here follows that presented by Flierl *et al.* (1981).

Suppose that there exists a steadily-travelling solution to (5.1) in the form $\psi = \varphi_s(\xi, \zeta)$ where $\xi = x - ct$ and $\zeta = y - \tilde{c}t$. Substitution into (5.1) yields

$$\partial(\varphi_s + c\zeta - \tilde{c}\xi, \Delta\varphi_s - \varphi_s + \zeta) = -\tilde{c}, \qquad (5.104)$$

where it is understood that $\partial(A, B) \equiv A_\xi B_\zeta - A_\zeta B_\xi$ and $\Delta \equiv \partial_{\xi\xi} + \partial_{\zeta\zeta}$. Suppose that there exists a region in the co-moving frame, denoted $\tilde{\Omega}$, for which $\partial\tilde{\Omega}$ corresponds to a *streak line*, i.e., a curve for which $\varphi_s + c\zeta - \tilde{c}\xi$ is constant. If (5.104) is integrated over $\tilde{\Omega}$, we obtain

$$\tilde{c}A_{\tilde{\Omega}} = -\iint\limits_{\tilde{\Omega}} \partial(\varphi_s + c\zeta - \tilde{c}\xi, \Delta\varphi_s - \varphi_s + \zeta) \, d\xi d\zeta$$

$$= -\iint\limits_{\tilde{\Omega}} \nabla \cdot [(\Delta\varphi_s - \varphi_s + \zeta) \, \mathbf{e}_3 \times \nabla(\varphi_s + c\zeta - \tilde{c}\xi)] \, d\xi d\zeta$$

$$= \int\limits_{\partial\tilde{\Omega}} (\Delta\varphi_s - \varphi_s + \zeta) \frac{\partial(\varphi_s + c\zeta - \tilde{c}\xi)}{\partial s} \, ds = 0,$$

since

$$\left[\frac{\partial(\varphi_s + c\zeta - \tilde{c}\xi)}{\partial s} \right]_{\partial\tilde{\Omega}} \equiv 0,$$

and where $A_{\tilde{\Omega}}$ is the area of the region $\tilde{\Omega}$. Hence we conclude that $\tilde{c}A_{\tilde{\Omega}} = 0$ which implies, assuming the nontrivial situation $A_{\tilde{\Omega}} \neq 0$, that $\tilde{c} = 0$ for these solutions. As a result, we shall henceforth restrict attention to steadily-travelling solutions of (5.1) in the form (5.101).

In light of the fact that the steadily travelling solutions we shall examine propagate only in the x-direction, we shall, for the remainder of this section, restrict the domain to either the entire plane or the channel given, respectively, by

$$\Omega = \mathbb{R}^2, \tag{5.105}$$

$$\Omega = \{(x,y) \mid -x_o < x < x_o,\; 0 < y < L\}. \tag{5.106}$$

If the domain is $\mathbb{R}^2$, we shall assume, unless otherwise stated, that the stream function approaches zero sufficiently rapidly and smoothly so that the integrated energy and enstrophy associated with the flow is bounded, i.e.,

$$\iint_{\mathbb{R}^2} \nabla\psi \cdot \nabla\psi + \psi^2 \; dxdy < \infty, \tag{5.107}$$

$$\iint_{\mathbb{R}^2} (\triangle\psi - \psi)^2 \; dxdy < \infty, \tag{5.108}$$

respectively. Under these assumptions, the Hamiltonian $H(q)$ for the *CHM* equation for the domain $\mathbb{R}^2$, can be expressed simply as

$$H(q) = \frac{1}{2} \iint_{\mathbb{R}^2} \nabla\psi \cdot \nabla\psi + \psi^2 \; dxdy$$

$$= -\frac{1}{2} \iint_{\mathbb{R}^2} (\triangle\psi - \psi)\,\psi \; dxdy$$

$$= -\frac{1}{2} \iint_{\mathbb{R}^2} (q - y)\,\psi \; dxdy. \tag{5.109}$$

If Ω is given by the channel domain (5.106), it will be assumed, unless otherwise stated, that the stream function satisfies, without loss of generality, the Dirichlet conditions

$$\psi(x, L, t) = \psi_L, \tag{5.110}$$

$$\psi(x, 0, t) = 0, \tag{5.111}$$

and, additionally, it will be assumed that $\psi(x, y, t)$ is smoothly periodic at $x = \pm x_o$. Under these boundary and periodicity conditions, the Hamiltonian $H(q)$ for the *CHM* equation for the periodic channel domain (5.106), can be expressed in the form

$$H(q) = \frac{1}{2} \iint\limits_{\Omega} \nabla\psi \cdot \nabla\psi + \psi^2 \, dx dy$$

$$-\psi_L \int_{-x_o}^{x_o} \psi_y(x, L, t) \, dx. \tag{5.112}$$

A Rossby wave solution

One example of a periodic steadily-travelling solution to (5.1) in the periodic channel (5.106) with the boundary conditions (5.110) and (5.111) is given by

$$\varphi_s(x - ct, y) = (\psi_L) y$$

$$+A \sin\left(\frac{m\pi y}{L}\right) \cos\left[\frac{n\pi}{x_o}(x - ct) + \phi\right], \tag{5.113}$$

where

$$c = -\frac{1 + \psi_L\left[\left(\frac{n\pi}{x_o}\right)^2 + \left(\frac{m\pi}{L}\right)^2\right]}{1 + \left(\frac{n\pi}{x_o}\right)^2 + \left(\frac{m\pi}{L}\right)^2}, \tag{5.114}$$

for arbitrary constant amplitude A and phase shift ϕ, with $m = 1, 2, 3, ...$, and $n = 0, 1, 2, ...$. This solution can be interpreted as the linear superposition of a constant flow in the x-direction and a Rossby wave satisfying the dispersion relation (5.4) with $c \equiv \omega/k$ with a constant mean flow included. Not that it is not possible to eliminate the term proportional to ψ_L in (5.114) by a transformation to an appropriately moving reference frame. This follows since the *CHM* equation is not invariant with respect to Galilean transformations.

It is straightforward to verify that the Rossby wave given by (5.113) and (5.114) satisfies the streak line-potential vorticity relationship

$$\varphi_s + cy = -\frac{\Delta\varphi_s - \varphi_s + y}{1 + \left(\frac{n\pi}{x_o}\right)^2 + \left(\frac{m\pi}{L}\right)^2}.$$

Thus, for this solution, the function $\Phi(q)$ in (5.103) is given by the linear relation

$$\Phi(q) = -\frac{q}{1 + \left(\frac{n\pi}{x_o}\right)^2 + \left(\frac{m\pi}{L}\right)^2}. \tag{5.115}$$

The *modon* solution

Modons are steadily-travelling dipole vortex solutions of the *CHM* equation which satisfy the finite energy and enstrophy constraints (5.107) and (5.108). Stern (1975) obtained the first steadily-travelling dipole vortex solution of the *CHM* equation in the rigid-lid limit where the stretching term ψ_t in (5.1) is neglected in comparison to the relative vorticity term $\Delta\psi_t$. The solution found by Stern had the undesirable property that the vorticity was not continuous everywhere in the spatial domain. Larichev & Reznik (1976) independently found a generalization of the Stern modon solution to the *CHM* equation (5.1) which had a continuous vorticity field everywhere in the spatial domain. The term *modon* was originally coined by Stern (1975) as a pun on the then ongoing joint *U.S.S.R.-U.S.A. Polymode* oceanographic experiment examining the evolution of oceanic eddies in the Atlantic Ocean.

It was originally thought that modons might correspond to a two dimensional soliton and, indeed, the title of the Larichev & Reznik (1976) paper *Two-dimensional Rossby soliton: An exact solution* suggests this point of view. This point of view was further supported by early numerical simulations which suggested that modons possessed soliton-like stability and interaction characteristics (e.g., McWilliams *et al.*, 1980, McWilliams & Zabusky, 1982).

The *CHM* equation does not, however, possess an infinite number of independent conservation laws so it is unlikely that there is a multidimensional inverse scattering procedure for constructing solutions to it. Consequently, if the conjecture is true that the existence of an inverse scattering procedure and a soliton solution to a given nonlinear partial differential equation are equivalent, then it is doubtful that modons are true multidimensional analogues of one dimensional solitons. Nevertheless, these solutions have been found for a number of fluid and plasma models (see, for example, the discussion in Swaters, 1989) and are thought to be important for describing aspects of the dynamics of coherent eddy-like features observed in turbulent flows.

For those streak lines which extend to infinity, it follows from (5.103) that

$$\Phi(y) = cy. \tag{5.116}$$

This follows since Φ is constant on each streak line and $\varphi_s \to 0$ as $r \equiv (x^2 + y^2)^{\frac{1}{2}} \to \infty$ for finite area-integrated energy and enstrophy solutions. Hence, for those streak lines which extend to infinity

$$\Phi(q_s) = cq_s, \tag{5.117}$$

which, when substituted into (5.103), implies that $\varphi_s(\xi, y)$ satisfies the homogeneous Helmholtz equation

$$\Delta\varphi_s - \left(1 + \frac{1}{c}\right)\varphi_s = 0. \tag{5.118}$$

For those streak lines that do extend to infinity there is no boundary condition which we can use to *a priori* determine the form of $\Phi(q_s)$. The modon solution is obtained by invoking the *ansatz* that

$$\Phi(q_s) = -\left(\kappa^2 + 1\right)^{-1} q_s, \qquad (5.119)$$

for all those streak lines which do not extend to infinity. The parameter κ is called the modon wavenumber and its determination is part of the solution. If (5.119) is substituted into (5.103), it follows that all those streak lines which do not extend to infinity satisfy the inhomogeneous Helmholtz equation

$$\Delta\varphi_s + \kappa^2\varphi_s = -\left(1 + c + c\kappa^2\right) y. \qquad (5.120)$$

The boundary between the region containing streak lines which extend to infinity and the region containing is assumed to be the circle $r = a$, where the parameter a is called the modon radius. Thus we may write

$$\Phi(q) = \begin{cases} cq, & \text{for } r > a, \\ -\left(\kappa^2 + 1\right)^{-1} q, & \text{for } r < a. \end{cases} \qquad (5.121)$$

The region $r > a$ will be referred to as the exterior region and the region $r < a$ will be referred to as the interior region.

We have to determine matching conditions for the solution on the boundary between the exterior and interior regions. Assuming a twice-continuously differentiable solution everywhere in the plane, it follows that

$$\lim_{r\downarrow a}\varphi_s + cy = \lim_{r\uparrow a}\varphi_s + cy,$$

which together with (5.103) and (5.121) implies that

$$\left[c + \left(\kappa^2 + 1\right)^{-1}\right] q_s|_{r=a} = 0.$$

Thus either $c = -\left(\kappa^2 + 1\right)^{-1}$ or $q_s|_{r=a} = 0$. If $c = -\left(\kappa^2 + 1\right)^{-1}$, then

$$1 + \frac{1}{c} = -\kappa^2 < 0.$$

But if $1 + c^{-1} < 0$, the solutions to (5.118) are given in terms of ordinary Bessel functions. Ordinary Bessel functions decay like $O\left(r^{-\frac{1}{2}}\right)$ as $r \equiv (x^2 + y^2)^{\frac{1}{2}} \to \infty$ which is not rapid enough for the area-integrated energy and enstrophy to be finite and thus the possibility that $c = \left(\kappa^2 + 1\right)^{-1}$ must be excluded. Hence we conclude that $q_s|_{r=a} = 0$, which in light of (5.103) implies

$$[\varphi_s + cy]_{r=a} = 0. \qquad (5.122)$$

This condition is necessary but not sufficient to ensure that q_s is continuous at $r = a$. In addition to (5.122) we will require that

$$\lim_{r\downarrow a} \nabla \varphi_s = \lim_{r\uparrow a} \nabla \varphi_s. \tag{5.123}$$

The condition that the area-integrated energy and enstrophy must be finite restricts the allowed values for the translation velocity c. We have already seen that if $1 + c^{-1} < 0$, there are no finite area-integrated energy and enstrophy solutions. If $c = -1$, then φ_s is harmonic and the relevant exterior region solutions are proportional to $r^{-n} [\cos(\theta), \sin(\theta)]$ for $n = 0, 1, 2, \ldots$ where $\theta = \tan^{-1}(y/\xi)$. Only the $r^{-1} \sin(\theta)$ mode is acceptable due to (5.122). However, this mode will not have finite area-integrated energy and enstrophy and the possibility that $c = -1$ must be excluded as well. Thus we conclude that only $1 + c^{-1} > 0$ is possible. This can be rearranged to imply that the set allowed translation velocities is

$$c < -1 \quad \text{or} \quad c > 0. \tag{5.124}$$

The set of allowed values for the modon translation velocity is disjoint from the set of allowed values for the x-direction phase velocity of Rossby waves. From (5.4) we see that the x-direction phase velocity of Rossby waves, denoted by c_{Rossby}, given by

$$c_{Rossby} \equiv \frac{\omega}{k} = \frac{-1}{l^2 + k^2 + 1},$$

implies

$$c_{Rossby} \in (-1, 0).$$

Assuming $1 + c^{-1} > 0$, the solution to the exterior problem (5.118) satisfying (5.122) which decays to zero at infinity is given by

$$\varphi_s(r, \theta) = -\frac{ac K_1(\gamma r/a) \sin(\theta)}{K_1(\gamma)}, \tag{5.125}$$

where $\gamma \equiv a(1 + c^{-1})^{\frac{1}{2}}$, and where $K_1(*)$ is the modified Bessel function of the first kind of order one.

The solution to the interior problem (5.120) satisfying (5.122) which is non-singular at the origin is given by

$$\varphi_s(r, \theta) = \frac{a(1 + c) J_1(\nu r/a) \sin(\theta)}{\kappa^2 J_1(\nu)}$$

$$-\frac{(1 + c + c\kappa^2) r \sin(\theta)}{\kappa^2}, \tag{5.126}$$

where $\nu \equiv \kappa a$, and where $J_1(*)$ is the ordinary Bessel function of the first kind of order one.

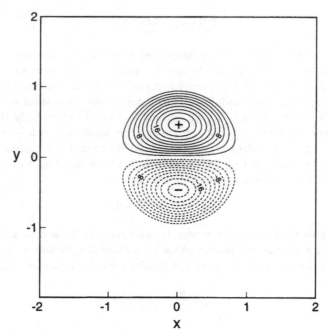

Figure 5.2. Contour plot of the relative vorticity $\Delta\varphi_s$ for the rightward-travelling modon.

The remaining constraint on the solution is (5.123). It is straightforward to verify that $\dfrac{\partial\varphi_s}{\partial\theta}$ is continuous on $r = a$. The condition that

$$\lim_{r\downarrow a}\frac{\partial\varphi_s}{\partial r} = \lim_{r\uparrow a}\frac{\partial\varphi_s}{\partial r},$$

implies

$$\frac{\gamma K_1(\gamma)}{K_2(\gamma)} = -\frac{\nu J_1(\nu)}{J_2(\nu)}. \tag{5.127}$$

This relation is called the modon dispersion relation. It is usual to consider that (5.127) defines $\kappa = \kappa(a, c)$. There are a countable infinity of $\kappa > 0$ solutions for each (a, c). The smallest nontrivial solution is called the ground state wavenumber and the corresponding modon the ground state modon. The ground state modon possess a single nodal line in the interior region along $x = 0.0$. The higher states have increasing nodal lines in the interior region.

 A contour graph of the relative vorticity field $\Delta\varphi_s$ for the ground state rightward-travelling modon with $a = c = 1.0$ and $\kappa \simeq 3.984$ (as determined by (5.127)) is presented in Figs. 5.2. The contour interval is ± 2.0 (without the 0$-$contour). We see an intense region of positive vorticity in the $y > 0$ region with a corresponding intense region of negative vorticity in the $y < 0$ region. The vorticity (and the stream function) exponentially decay to zero for large distances from its center. The entire structure moves from left to right because the $y > 0$ vortex core acts to advect the $y < 0$ vortex core in that direction and vice versa. Thus, there is a net impulse in the positive x direction.

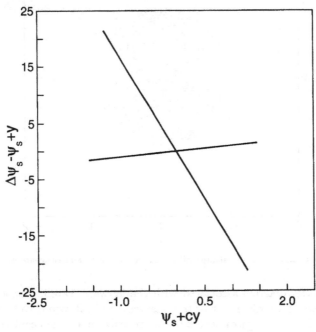

Figure 5.3. Scatter diagram of $(\varphi_s + cy, \Delta\varphi_s - \varphi_s + y)$ for the rightward-travelling modon.

In Fig. 5.3 we present the scatter diagram, i.e., the graph of

$$(\varphi_s + cy, \ (\Delta - 1)\,\varphi_s + y)\,,$$

for the rightward-travelling modon shown in Fig. 5.2. Note the nonanalytic structure of the graph. As we discuss more fully momentarily, the nonanalytic structure of $\Phi(q)$ makes the construction of a variational principle and stability theory for the modon problematic.

 In Fig. 5.4 we present a contour graph of the relative vorticity $\Delta\varphi_s$ for the ground state leftward-travelling modon with $a = 1.0$, $c = -2.0$ and $\kappa \simeq 3.883$. The

contour interval is ±3.0 with the 0—contour deleted. The leftward-travelling modon
has the intense positive vorticity region in the $y < 0$ half-plane with a corresponding
region of intense negative vorticity in the $y > 0$ half-plane. The overall advection
results in a net impulse in the negative x direction and thus right to left motion.

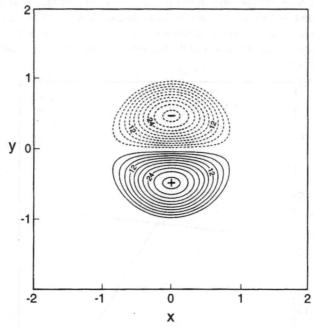

Figure 5.4. Contour plot of the relative vorticity $\Delta\varphi_s$ for the leftward-travelling modon.

The scatter diagram associated with the leftward-travelling modon is shown
in Fig. 5.5. The scatter diagrams for the rightward and leftward travelling modon are
similar. Note that both figures have a line segment with a large negative slope. This
component is associated with the modon field in the interior region $r < a$. The second
line segment in the scatter diagrams corresponds to the modon field in the exterior
region $r > a$. (The origin in the scatter diagrams corresponds to the modon boundary
$r = a$.) Thus in the scatter diagram associated with the rightward-travelling modon,
i.e., Fig. 5.3, the line segment associated with the exterior region has a positive
slope. Whereas for the leftward-travelling modon, the line segment associated with
the exterior region has a negative slope.

This is an important point to note since it implies that $\Phi'(q_s)$, to the extent
that it is defined, for the modon has the same sign everywhere for the leftward-
travelling solution and changes sign in the rightward-travelling solution. The strategy
associated with our stability argument revolves around establishing conditions for the
definiteness of the second variation of the constrained Hamiltonian (or some other

functional). This procedure usually involved assuming that $\Phi'(q_s)$, or its equivalent, is definite. Thus we see, at least for the rightward-travelling modon, that this line of argument will fail.

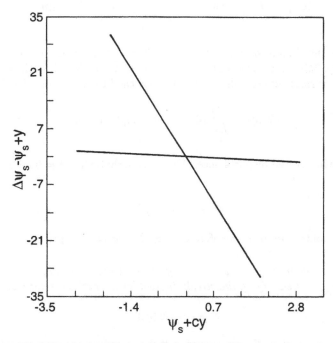

Figure 5.5. Scatter diagram of $(\varphi_s + cy, \Delta\varphi_s - \varphi_s + y)$ for the leftward-travelling modon.

Since the modon boundary is a streak line in the co-moving frame of reference, the fluid in the modon interior is isolated from the modon exterior. In the original (x, y) stationary reference frame the fluid in the interior region is therefore trapped and is actually transported with the moving modon. These solutions provide a model for the transport of fluids parcels from one region of the spatial domain to another. Finally, we point out that Kloeden (1987) has presented a proof for the assertion that the modon is the *unique* steadily-travelling solution (with finite area-integrated energy and enstrophy) to the *CHM* equation on the domain $\mathbb{R}^2$.

Variational principles for general steadily-travelling solutions

Benjamin (1984) gave the following variational principle for steadily-travelling solutions to the *CHM* equation. Consider the constrained Hamiltonian functional given by

$$\mathcal{H}(q) = H(q) - cM(q) + \iint_{\Omega} \left\{ \int_{y}^{q} \Phi(\xi) d\xi \right\} dx dy,$$

where $H(q)$ is the Hamiltonian (5.109) or (5.112) and $M(q)$ is the impulse functional (5.41). Clearly, $\mathcal{H}(q)$ is an invariant of the *CHM* equation since it is the sum of three individually invariant functionals. If we compute the first variation $\delta \mathcal{H}(q)$, we find

$$\delta \mathcal{H}(q) = \iint_{\Omega} [\Phi(q) - \psi - cy] \, \delta q \, dx dy,$$

where we have integrated by parts once. Thus we see that $\delta \mathcal{H}(q_s) = 0$ for all variations δq if

$$\varphi_s + cy = \Phi(q_s),$$

i.e., if (5.103) holds. We have therefore proven the variational principle

Theorem 5.21 *Suppose $\varphi_s = \varphi_s(x - ct, y)$ is a steadily-travelling solution to the CHM equation (5.1) satisfying the streak line-potential vorticity relationship*

$$\varphi_s + cy = \Phi(q_s),$$

where $q_s \equiv \Delta \varphi_s - \varphi_s + y$, then φ_s satisfies the first order necessary condition for an extremal of the constrained Hamiltonian

$$\mathcal{H}(q) = H(q) - cM(q) + \iint_{\Omega} \left\{ \int_{y}^{q} \Phi(\xi) d\xi \right\} dx dy, \qquad (5.128)$$

i.e., $\delta \mathcal{H}(q_s) = 0$, where $H(q)$ is the Hamiltonian (5.109) or (5.112) and $M(q)$ is the impulse functional (5.41).

It is also possible to give an alternate variational principle (Benjamin, 1984) for steadily-travelling solutions of the form (5.103). We will examine the two domains $\mathbb{R}^2$ and the periodic channel separately.

Consider first the domain $\mathbb{R}^2$ and the functional $\widetilde{\mathcal{H}}(q)$ given by

$$\widetilde{\mathcal{H}}(q) = \iint_{\mathbb{R}^2} \left\{ \frac{1}{2} \left[|\nabla(\psi + cy)|^2 + (\psi + cy)^2 - c^2(1 + y^2) \right] \right.$$

$$+ \int\limits_{y}^{q} \Phi(\xi)d\xi \Bigg\} \; dxdy, \tag{5.129}$$

where we are assuming that $|\psi| \to 0$ as $x^2 + y^2 \to \infty$ sufficiently rapidly so that the area-integrated energy and enstrophy is finite. The term $c^2(1 + y^2)$ in the integrand of $\widetilde{\mathcal{H}}(q)$ is required so that

$$|\nabla(\psi + cy)|^2 + (\psi + cy)^2 - c^2(1 + y^2) \to 0,$$

as $x^2 + y^2 \to \infty$, which is necessary for the integral to exist.

The first thing that needs to be shown is that $\widetilde{\mathcal{H}}(q)$ is an invariant for the *CHM* equation (5.1). Since the Casimir part of (5.129) is trivially invariant, it suffices to show the invariance of the non-Casimir part only, given by,

$$\widetilde{\mathcal{H}}_{nc}(q) \equiv \frac{1}{2} \iint\limits_{\mathbb{R}^2} |\nabla(\psi + cy)|^2 + (\psi + cy)^2 - c^2(1 + y^2) \; dxdy.$$

It follows that

$$\frac{d\widetilde{\mathcal{H}}_{nc}(q)}{dt} = - \iint\limits_{\mathbb{R}^2} (\psi + cy)(\Delta\psi - \psi)_t \; dxdy$$

$$= - \iint\limits_{\mathbb{R}^2} \psi(\Delta\psi - \psi)_t \; dxdy - c \iint\limits_{\mathbb{R}^2} y(\Delta\psi - \psi)_t \; dxdy.$$

Now

$$\iint\limits_{\mathbb{R}^2} \psi(\Delta\psi - \psi)_t \; dxdy = 0,$$

is the statement that energy is conserved, and

$$\iint\limits_{\mathbb{R}^2} y(\Delta\psi - \psi)_t \; dxdy = 0,$$

is the statement that the x-direction impulse is an invariant (which we showed as a consequence of Noether's Theorem; see (5.41)). Hence we conclude that

$$\frac{d\widetilde{\mathcal{H}}(q)}{dt} = 0.$$

The first variation of $\widetilde{\mathcal{H}}(q)$ is given by

$$\delta\widetilde{\mathcal{H}}(q) = \iint_{\Omega} [\Phi(q) - \psi - cy]\,\delta q\,dxdy,$$

from which it follows that $\delta\widetilde{\mathcal{H}}(q_s) = 0$ provided that (5.103) holds. We have therefore established the alternate variational principle

Theorem 5.22 *Suppose* $\varphi_s = \varphi_s(x - ct, y)$ *is a steadily-travelling solution to the CHM equation (5.1) on the spatial domain is* $\mathbb{R}^2$ *(with the boundary condition* $|\psi| \to 0$ *as* $x^2 + y^2 \to \infty$ *sufficiently rapidly so that the area-integrated energy and enstrophy is finite), satisfying the streak line-potential vorticity relationship*

$$\varphi_s + cy = \Phi(q_s),$$

where $q_s \equiv \Delta\varphi_s - \varphi_s + y$, *then* φ_s *satisfies the first order necessary condition, i.e.,* $\delta\widetilde{\mathcal{H}}(q_s) = 0$, *for an extremal of the conserved functional*

$$\widetilde{\mathcal{H}}(q) = \iint_{\mathbb{R}^2} \left\{ \frac{1}{2}\left[|\nabla(\psi + cy)|^2 + (\psi + cy)^2 - c^2(1 + y^2)\right] \right.$$

$$\left. + \int_y^q \Phi(\xi)d\xi \right\}\,dxdy.$$

Let us now consider the periodic channel domain Ω given by (5.106) with the boundary conditions (5.110) and (5.111) where it is assumed that ψ is smoothly periodic at $x = \pm x_o$, and the functional

$$\widehat{\mathcal{H}}(q) = \iint_{\Omega} \left\{ \frac{1}{2}\left[|\nabla(\psi + cy)|^2 + (\psi + cy)^2\right] \right.$$

$$\left. + \int_y^q \Phi(\xi)d\xi \right\}\,dxdy - (\varphi_L + cL)\int_{-x_o}^{x_o} \psi_y(x, L, t)\,dx. \tag{5.130}$$

The demonstration that $\widehat{\mathcal{H}}(q)$ is an invariant is similar to that given for $\widetilde{\mathcal{H}}(q)$ and left as an exercise for the reader. The first variation of $\widehat{\mathcal{H}}(q)$ is given by

$$\delta\widehat{\mathcal{H}}(q) = \iint_{\Omega} [\Phi(q) - \psi - cy]\,\delta q\,dxdy,$$

where we have integrated by parts once. It follows that $\delta\widehat{\mathcal{H}}(q_s) = 0$ provided that (5.103) holds. Thus we have established the alternate variational principle

Theorem 5.23 *Let $\varphi_s = \varphi_s(x - ct, y)$ be a steadily-travelling solution of the CHM equation (5.1) in the periodic channel domain (5.106) satisfying the boundary conditions (5.110) and (5.111) and the smooth periodicity conditions at $x = \pm x_o$, then $\delta \widehat{\mathcal{H}}(q_s) = 0$ for $\widehat{\mathcal{H}}(q)$ given by*

$$\widehat{\mathcal{H}}(q) = \iint\limits_{\Omega} \left\{ \frac{1}{2} \left[|\nabla(\psi + cy)|^2 + (\psi + cy)^2 \right] \right.$$

$$\left. + \int\limits_y^q \Phi(\xi) d\xi \right\} \; dxdy - (\varphi_L + cL) \int\limits_{-x_o}^{x_o} \psi_y(x, L, t) \; dx.$$

With more than one variational principle, there is perhaps a degree of ambiguity as to which principle to use. Benjamin (1984) suggested that variational principles based on (5.129) and (5.130) might be particularly useful in developing a mathematical existence theory for steadily-travelling solutions to the *CHM* equation. On the other hand, since the variational principle based on (5.128) is very much similar to soliton variational principles (to be discussed in Chapter 6 in relation to the *KdV* equation), and these principles provide the mathematical framework for soliton stability proofs, Benjamin (1984) suggested that the variational principle based on (5.128) might be more appropriate in determining the stability characteristics of the steadily-travelling solutions to the *CHM* equation.

Stability theory for steadily-travelling solutions

It is not, however, possible to establish the linear let alone the nonlinear stability of *steadily-travelling* solutions to the *CHM* equations via Arnol'd-like stability arguments as developed in Section 5.4. That is, it is not possible to show that the second variation of the functionals we have used for the variational principles is positive definite for all perturbations. This is a consequence of an appropriately extended form of Andrews's Theorem applicable to steadily-travelling solutions of the *CHM* equations in the domains $\Omega = \mathbb{R}^2$ or the periodic channel. To see the breakdown let us, for example, examine the second variation of $\mathcal{H}(q)$ given in the variational principle Theorem 5.21.

The second variation of $\mathcal{H}(q)$ evaluated at the steadily-travelling solution (5.103) can be written in the form

$$\delta^2 \mathcal{H}(q_s) = \iint\limits_{\Omega} \left\{ |\nabla \delta \varphi|^2 + (\delta \varphi)^2 + \Phi_s'(\delta q)^2 \right\} \; dxdy, \qquad (5.131)$$

where

$$\Phi_s' \equiv \left[\frac{d\Phi(q)}{dq} \right]_{q=q_s}. \qquad (5.132)$$

It is straightforward to verify that $\delta^2\widetilde{\mathcal{H}}(q_s)$ and $\delta^2\widehat{\mathcal{H}}(q_s)$ are identical to (5.131). Thus the following remarks are applicable to these functionals as well.

We wish to show that $\delta^2\mathcal{H}(q_s)$ is an invariant of the linear stability equation associated with the steadily-travelling solution $\varphi_s(\xi, y)$. Suppose that the total stream function $\psi(x, y, t)$ can be written in the form

$$\psi = \varphi_s(x - ct, y) + \delta\varphi(x, y, t), \tag{5.133}$$

where $\varphi_s(x - ct, y)$ satisfies (5.103) and where $\delta\varphi(x, y, t)$ is the perturbation stream function.. Substitution of (5.133) into (5.1) yields, after neglecting the quadratic perturbation terms,

$$(\Delta\delta\varphi - \delta\varphi)_t + \partial(\delta\varphi, q_s) + \partial(\varphi_s, \Delta\delta\varphi - \delta\varphi) = 0.$$

If (5.103) is used we may write this equation in the form

$$(\partial_t + c\partial_x)\,\delta q + \partial\,(q_s, \Phi_s'\delta q - \delta\varphi) = 0. \tag{5.134}$$

The linear stability equation (5.134) is identical in appearance to (5.50) except that the local time rate of change in (5.50) is replaced by the Lagrangian time rate of change moving with the steadily-travelling solution.

Theorem 5.24 $\delta^2\mathcal{H}(q_s)$ *is an invariant of the linear stability equation (5.134).*

Proof. We proceed directly

$$\frac{\partial\delta^2\mathcal{H}(q_s)}{\partial t} = 2\iint\limits_{\Omega} \left\{ (\Phi_s'\delta q - \delta\varphi)\,\delta q_t - \frac{c}{2}\Phi_s''\,(\delta q)^2\,q_{s_x} \right\}\,dx\,dy \tag{5.135}$$

where we have integrated by parts once exploiting the boundary condition $\delta\varphi|_{\partial\Omega} = 0$, substituted $q_{s_t} \equiv -cq_{s_x}$, and where

$$\Phi_s'' \equiv \left[\frac{d^2\Phi(q)}{dq^2}\right]_{q=q_s} \tag{5.136}$$

The reader is reminded that the domain Ω in this section is assumed to be either $\mathbb{R}^2$ or the periodic channel (5.106) (with the appropriate boundary conditions as specified earlier). If the linear stability equation (5.134) is used to eliminate δq_t, the first term in the integrand of (5.135) can be re-written in the form

$$(\Phi_s'\delta q - \delta\varphi)\,\delta q_t = -\frac{1}{2}\partial\,\left(q_s, [\Phi_s'\delta q - \delta\varphi]^2\right)$$

$$-c\,(\Phi_s'\delta q - \delta\varphi)\,\delta q_x$$

$$= -\frac{1}{2}\partial \left(\Delta\varphi_s - \varphi_s, [\Phi'_s \delta q - \delta\varphi]^2 \right)$$

$$+\frac{1}{2}\left([\Phi'_s \delta q - \delta\varphi]^2 \right)_x$$

$$+\frac{c}{2}\Phi''_s (\delta q)^2 q_{s_x} + c\nabla \cdot (\delta\varphi \nabla \delta\varphi_x)$$

$$-\frac{c}{2}\left(\Phi'_s (\delta q)^2 + (\delta\varphi)^2 + \nabla\delta\varphi \cdot \nabla\delta\varphi \right)_x.$$

If this expression is substituted into (5.135), we obtain

$$\frac{\partial \delta^2 \mathcal{H}(q_s)}{\partial t} = \iint_\Omega \left\{ -\partial \left(\Delta\varphi_s - \varphi_s, [\Phi'_s \delta q - \delta\varphi]^2 \right) \right.$$

$$+\left([\Phi'_s \delta q - \delta\varphi]^2 \right)_x + 2c\nabla \cdot (\delta\varphi \nabla \delta\varphi_x)$$

$$\left. -c\left(\Phi'_s (\delta q)^2 + (\delta\varphi)^2 + \nabla\delta\varphi \cdot \nabla\delta\varphi \right)_x \right\} dxdy.$$

The non-Jacobian terms all integrate to zero if $\Omega = \mathbb{R}^2$ because $|\delta\varphi| \to 0$ and, from (5.116), $\Phi'_s \to c$ as $x^2 + y^2 \to \infty$. If Ω is the periodic domain (5.106), all the non-Jacobian terms integrate to zero as well because of the periodicity at $x = \pm x_o$ and the fact that $\delta\varphi = 0$ on $y = 0$ and L, respectively. Thus, it follows that

$$\frac{\partial \delta^2 \mathcal{H}(q_s)}{\partial t} = -\iint_\Omega \partial \left(\Delta\varphi_s - \varphi_s, [\Phi'_s \delta q - \delta\varphi]^2 \right) dxdy$$

$$= -\iint_\Omega \nabla \cdot \left[\mathbf{e}_3 \times \nabla (\Delta\varphi_s - \varphi_s) (\Phi'_s \delta q - \delta\varphi)^2 \right] dxdy$$

$$= \int_{\partial\Omega} \frac{\partial (\Delta\varphi_s - \varphi_s)}{\partial s} (\Phi'_s \delta q - \delta\varphi)^2 ds$$

$$= \int_{\partial\Omega} \frac{\partial (\Delta\varphi_s - \varphi_s)}{\partial s} (\Phi'_s \delta q - \delta\varphi)^2 ds. \tag{5.137}$$

If $\Omega = \mathbb{R}^2$, we may write this integral as

$$\frac{\partial \delta^2 \mathcal{H}(q_s)}{\partial t} = \lim_{r \to \infty} \int_0^{2\pi} \frac{\partial \left(\Delta \varphi_s - \varphi_s \right)}{\partial \theta} \left(\Phi_s' \delta q - \delta \varphi \right)^2 d\theta,$$

where

$$r \equiv \left(x^2 + y^2 \right)^{\frac{1}{2}} \text{ and } \theta \equiv \tan^{-1} \left(\frac{y}{x} \right).$$

However, this integral vanishes since all terms in the integrand vanish as $x^2 + y^2 \to \infty$. If Ω is the periodic channel, then (5.137) can be written in the form

$$\int_{\partial \Omega} \frac{\partial q_s}{\partial s} \left(\Phi_s' \delta q - \delta \varphi \right)^2 ds =$$

$$\int_{-x_o}^{x_o} \left[\frac{\partial \left(\Delta \varphi_s - \varphi_s \right)}{\partial x} \left(\Phi_s' \delta q \right)^2 \right]_{y=0} dx$$

$$+ \int_{-x_o}^{x_o} \left[\frac{\partial \left(\Delta \varphi_s - \varphi_s \right)}{\partial x} \left(\Phi_s' \delta q \right)^2 \right]_{y=L} dx, \qquad (5.138)$$

where we have used the fact that $[\delta \varphi]_{y=0,L} \equiv 0$. It follows from (5.103) that

$$\frac{\partial \varphi_s}{\partial x} = \Phi_s' \frac{\partial \left(\Delta \varphi_s - \varphi_s \right)}{\partial x}.$$

If this is substituted into (5.138), we obtain

$$\int_{\partial \Omega} \frac{\partial \left(\Delta \varphi_s - \varphi_s \right)}{\partial s} \left(\Phi_s' \delta q - \delta \varphi \right)^2 ds =$$

$$\int_{-x_o}^{x_o} \left[\frac{\partial \varphi_s}{\partial x} \Phi_s' \left(\delta q \right)^2 \right]_{y=0} dx$$

$$+ \int_{-x_o}^{x_o} \left[\frac{\partial \varphi_s}{\partial x} \Phi_s' \left(\delta q \right)^2 \right]_{y=L} dx \equiv 0,$$

since $\dfrac{\partial \varphi_s}{\partial x} \equiv 0$ on $y = 0$ and L, respectively. Hence we conclude $\delta^2 \mathcal{H}(q_s)$ is an invariant of the linear stability equation (5.134). ∎

In principle, the linear stability in the sense of Liapunov of $\varphi_s(x - ct, y)$ could be established if the function $\Phi_s(q_s)$ as determined by (5.103) satisfied

$$0 \le \Phi'_s \le \mu < \infty, \tag{5.139}$$

for some constant μ for all $(x, y, t) \in \mathbb{R}^2 \times [0, \infty)$. This would be just the analogue of Theorem 5.9 for the steadily-travelling solutions. However, and this is the crucial point, we now show

Theorem 5.25 *There are no nontrivial steadily-travelling solutions to (5.1) in the plane or in the periodic channel for which (5.139) holds.*

Proof. The proof is an application of Andrews's Theorem for the steadily-travelling solutions. If (5.103) is differentiated with respect to x, it follows that

$$\varphi_{s_x} = \Phi'_s(\Delta \varphi_{s_x} - \varphi_{s_x}).$$

If this expression is multiplied through by $(\Delta \varphi_{s_x} - \varphi_{s_x})$ and integrated over Ω we obtain

$$\iint\limits_{\Omega} (\Delta \varphi_{s_x} - \varphi_{s_x})^2 \Phi'_s + |\nabla \varphi_{s_x}|^2 + (\varphi_{s_x})^2 dx dy = 0, \tag{5.140}$$

since $|\varphi_s| \to 0$ as $x^2 + y^2 \to \infty$ if $\Omega = \mathbb{R}^2$, and since $\varphi_{s_x} = 0$ on $y = 0$ and L for the periodic channel. However, if (5.140) holds, then it necessarily follows that $\varphi_{s_x} \equiv 0$. If $\Omega = \mathbb{R}^2$, then the fact that $|\varphi_s| \to 0$ as $x^2 + y^2 \to \infty$ implies that $\varphi_s \equiv 0$ and if Ω is the periodic channel then $\varphi_s \equiv \varphi_s(y)$ and there is a trivial dependence on the phase $x - ct$. Therefore, there are no nontrivial steadily-travelling solutions to (5.1) that can satisfy the Arnol'd-like stability condition (5.139). ∎

For the periodic channel (5.106), there exists a Poincaré inequality for the perturbation stream function.. Thus, in principle, the linear stability in the sense of Liapunov could be established for the steadily-travelling solutions if there existed negative constants μ_1 and μ_2 for which

$$-\infty < \mu_1 \le \Phi'_s \le \mu_2 < -\dfrac{\lambda_{\min} + 1}{\lambda^2_{\min}} < 0, \tag{5.141}$$

for all $(x, y, t) \in \Omega \times [0, \infty)$, where $\lambda_{\min}$ is the strictly positive minimum eigenvalue of the boundary value problem

$$\Delta \phi + \lambda \phi = 0, \ (x, y) \in \Omega, \tag{5.142}$$

$$\phi|_{\partial \Omega} = 0. \tag{5.143}$$

This would be just the analogue of Theorem 5.12 for the steadily-travelling solutions. However, here again we have

Theorem 5.26 *There are no nontrivial steadily-travelling solutions to (5.1) in the periodic channel for which (5.141) holds. That is, if (5.141) holds, then* $\varphi_{s_x} \equiv 0$.

Proof. The proof is identical to the second half of the proof of Theorem 5.15 and is left as an exercise for the reader. ∎

These two theorems underscore a very important open research problem: developing a mathematical theory for determining the stability or instability of steadily-travelling solutions to (5.1).

Remarks on the stability theory of modons and Rossby waves

One example of a steadily-travelling solution is the modon. There have been many attempts to establish the linear stability of modons (e.g., Petviashvili, 1983; Pierini, 1985; Swaters, 1986; Laedke & Spatschek, 1986; and Sakuma & Ghil, 1990, 1991). All of these analyses are wrong (Carnevale *et al.*, 1988; Ripa, 1992; Nycander, 1992). The entire problem of rigorously establishing the stability, assuming it exists, of modons is, at the present time, an open research question.

For the leftward-travelling solutions, i.e., the $c < -1$ solutions, it would seem that stability is out of the question because of the so-called *tilt instability* (Makino *et al.*, 1981; Nycander, 1992). This instability corresponds to perturbing the direction of propagation of the leftward-travelling modon so that initially there is a small component in the y-direction. No matter how small this initial perturbation, if it is non-zero, the modon follows a very complicated cycloid-like trajectory which can only be identified as a large-amplitude deviation from the trajectory of an unperturbed leftward-travelling modon. The physical reason for the instability is that the ambient background vorticity gradient (i.e., the beta effect) couples with the vorticity signature of the leftward-travelling modon to exacerbate the deviation.

While leftward-travelling modons are thus demonstrably unstable, the situation for the rightward-travelling solutions, i.e., the $c > 0$ solutions, is not so clear. Numerical experiments (e.g., McWilliams *et al.*, 1981) seem to suggest that these rightward-travelling solutions are stable for a rather large class of perturbations. However it needs to be emphasized again that numerical simulations are one thing and mathematical proof quite another.

One can easily see what the mathematical difficulty is in attempting to use the functional $\delta^2 \mathcal{H}(q_s)$ to examine the stability of the modon. If Φ'_s is computed for the modon $\Phi(q)$ as given by (5.121) and substituted into (5.131), it follows that the functional

$$\mathrm{H}(\delta q) \equiv \iint_{\mathbb{R}^2} \nabla \delta\varphi \cdot \nabla \delta\varphi + (\delta\varphi)^2 \, dxdy$$

$$-\frac{1}{\kappa^2 + 1} \iint_{\xi^2 + y^2 < a^2} (\delta q)^2 \, dxdy + c \iint_{\xi^2 + y^2 > a^2} (\delta q)^2 \, dxdy, \qquad (5.144)$$

where $\xi \equiv x - ct$, is an invariant of the linear stability equation associated with the modon solution.

It is important to point out that because of the finite step discontinuity in the modon function $\Phi(q)$ at the modon boundary $\xi^2 + y^2 = a^2$, it will follow that the corresponding $\mathcal{H}(q)$ will *not* be an invariant of (5.1) for every $q(x, y, t)$ since, in general, $q|_{\xi^2 + y^2 = a^2} \neq 0$. However, it *is* true that

$$\frac{\partial \mathrm{H}(\delta q)}{\partial t} \equiv 0,$$

for all perturbations δq as determined by the linear stability equation for the modon. The details of this demonstration are left as an exercise for the reader. It is because $\mathcal{H}(q)$ will not be an invariant of (5.1) for every $q(x, y, t)$ for the modon $\Phi(q)$ relation that we have decided to label what otherwise would be $\delta^2 \mathcal{H}(q_s)$ as $\mathrm{H}(\delta q)$ for the purposes of this discussion.

We can immediately see the difficulty in attempting to show that $\mathrm{H}(\delta q)$ is either positive or negative definite for all $\delta\varphi$. Consider the case if $c < -1$ (i.e., the leftward-travelling solution). The integrals associated with the perturbation enstrophy are clearly negative definite. The Arnol'd-like argument would be to show that $\mathrm{H}(\delta q)$ is negative definite. However, this requires Poincaré inequalities of the form established in Lemma 5.10 and 5.11. But this is impossible for the exterior region. The exterior region is not bounded in any direction so a Poincaré inequality cannot hold.

In the case that $c > 0$ (i.e., the rightward-travelling solution), Φ'_s changes sign across the modon boundary. Since Φ'_s is not definite it obviously is unable to satisfy the conditions for Arnol'd stability. Thus, on the basis of fairly elementary arguments, it is not possible to establish the definiteness of $\mathrm{H}(\delta q)$ for either the leftward or rightward-travelling modon via a straightforward application of Arnol'd stability theory.

The breakdown of the theory for the Rossby wave is also easy to see. For the Rossby wave solution (5.113) in the periodic channel (5.106) with the boundary conditions (5.110) and (5.111), $\Phi(q)$ is given by (5.115). Thus, $\delta^2 \mathcal{H}(q_s)$ is given by

$$\delta^2 \mathcal{H}(q_s) = \iint\limits_{\Omega} \left\{ |\nabla \delta\varphi|^2 + (\delta\varphi)^2 \right.$$

$$\left. - \frac{(\delta q)^2}{1 + \left(\frac{n\pi}{x_o}\right)^2 + \left(\frac{m\pi}{L}\right)^2} \right\} dx dy. \tag{5.145}$$

Since $\Phi'_s < 0$, the Arnol'd-like stability argument would be to show that (5.141) holds.

But this can never hold since $\lambda_{\min} = \left(\frac{\pi}{L}\right)^2$ and

$$\frac{1}{1 + \left(\frac{n\pi}{x_o}\right)^2 + \left(\frac{m\pi}{L}\right)^2} < \frac{1}{\left(\frac{\pi}{L}\right)^2} < \frac{1 + \frac{\pi}{L}}{\left(\frac{\pi}{L}\right)^2},$$

which implies that

$$\Phi'_s \equiv -\frac{1}{1 + \left(\frac{n\pi}{x_o}\right)^2 + \left(\frac{m\pi}{L}\right)^2}$$

$$> -\frac{1 + \frac{\pi}{L}}{\left(\frac{\pi}{L}\right)^2} \equiv \frac{1 + \lambda_{\min}}{\lambda_{\min}^2}.$$

Hence, (5.141) can never hold for the Rossby wave (5.113).

As we have alluded to before, the failure of the "Arnol'd method", to establish stability for a particular flow or even class of flows, does not necessarily imply that these flows are in fact unstable. Although, for the cases just described, it is known that the Rossby wave *is* unstable (e.g., Lorenz, 1972) and it is not unreasonable to suspect that the modon (at least the $c < 0$ solution) is as well (e.g., Nycander, 1992).

5.6   Exercises

Exercise 5.1 *Show directly that the circulation integrals*

$$\oint_{\partial \Omega_i} \mathbf{n}_i \cdot \nabla \psi \, ds,$$

for $i = 1, ..., n$ are invariants of the CHM equation (5.1).

Exercise 5.2 *Show that (5.25) holds for the shallow-water equations (5.10), (5.11) and (5.12) by forming the vorticity equation*

$$\frac{\partial (5.11)}{\partial x} - \frac{\partial (5.10)}{\partial y},$$

and then eliminating the horizontal divergence of the velocity field terms using (5.12).

Exercise 5.3 *Show that the leading order nontrivial balance obtained after substituting the asymptotic expansion (5.13) into (5.25) is just the CHM equation (5.1).*

Exercise 5.4 *Show that the linearized CHM equation (i.e., (5.1) without the nonlinear $\partial\left(\psi, \Delta\psi\right)$ term), corresponds to the first order necessary condition for an extremal to the Lagrangian*

$$\mathcal{L} = \iint_{\Omega} \nabla\chi_t \cdot \nabla\chi_t + (\chi_t)^2 - \chi_t\chi_x \; dxdy,$$

with the Clebsch variable transformation $\psi = \chi_t$.

Exercise 5.5 *Prove Theorem 5.1.*

Exercise 5.6 *Consider the unbounded domain $\Omega = \mathbb{R}^2$, where it is assumed that $|\psi| \to 0$ sufficiently rapidly and smoothly so that the area-integrated energy and enstrophy are finite. Show that the Hamiltonian structure for the CHM equation is invariant for arbitrary translations with respect to y. Use Noether's Theorem to conclude that the y-direction linear momentum functional*

$$\iint_{\Omega} v \; dxdy,$$

is a constant of the motion.

Exercise 5.7 *Prove Lemma 5.13.*

Exercise 5.8 *Prove Theorem 5.14.*

Exercise 5.9 *Prove Lemma 5.16.*

Exercise 5.10 *Show that the functional*

$$\widehat{\mathcal{H}}\left(q\right) = \iint_{\Omega} \left\{ \frac{1}{2} \left[|\nabla(\psi + cy)|^2 + (\psi + cy)^2 \right] \right.$$

$$\left. + \int_{y}^{q} \Phi(\xi)d\xi \right\} \; dxdy - (\varphi_L + cL) \int_{-x_o}^{x_o} \psi_y\left(x, L, t\right) \; dx,$$

is an invariant of the CHM equation for the periodic channel domain (5.106) with boundary conditions (5.110) and (5.111).

Exercise 5.11 *Prove Theorem 5.26.*

Exercise 5.12 *Show that the functional*

$$H(\delta q) \equiv \iint\limits_{\mathbb{R}^2} \nabla \delta \varphi \cdot \nabla \delta \varphi + (\delta \varphi)^2 \, dx dy$$

$$-\frac{1}{\kappa^2 + 1} \iint\limits_{\xi^2 + y^2 < a^2} (\delta q)^2 \, dx dy + c \iint\limits_{\xi^2 + y^2 > a^2} (\delta q)^2 \, dx dy,$$

is an invariant of the linear stability equation associated with the modon.

Exercise 5.13 *Show that the functional* $\mathcal{H}(q)$ *given by (5.128) with the modon* $\Phi(q)$ *function can be written in the form*

$$\mathcal{H}(q) = \iint\limits_{\mathbb{R}^2} \left\{ \nabla \psi \cdot \nabla \psi + \psi^2 - cy(\Delta \psi - \psi) \right\} \, dx dy$$

$$-\frac{1}{\kappa^2 + 1} \iint\limits_{\xi^2 + y^2 < a^2} \left\{ (\Delta \psi - \psi + y)^2 - y^2 \right\} \, dx dy$$

$$+c \iint\limits_{\xi^2 + y^2 > a^2} \left\{ (\Delta \psi - \psi + y)^2 - y^2 \right\} \, dx dy,$$

and consequently show that $H(\delta q)$ *as given in the previous question is formally* $\delta^2 \mathcal{H}(q_{modon})$

Exercise 5.14 *Show that* $\mathcal{H}(q)$ *as given in the previous question is not, in general, an invariant of the CHM equation for arbitrary smooth solutions* $\psi(x, y, t)$.

Exercise 5.15 *Show that if*

$$[\Delta \psi - \psi + y]_{\xi^2 + y^2 = a^2} \equiv 0,$$

then

$$\frac{d\mathcal{H}(q)}{dt} = 0.$$

Chapter 6

The *KdV* equation

One of the early successes of Hamiltonian-based stability theory was Benjamin's (1972) (with mathematical corrections by Bona, 1975) proof of the nonlinear stability of the soliton solution to the *Korteweg-de Vries* or, more simply, the *KdV* equation. In this chapter we will describe the Hamiltonian structure of the *KdV* equation and present the stability theory as a consequence of the underlying Hamiltonian formalism.

There have been many excellent accounts of the *KdV* equation, its solutions and underlying mathematical structure. It is, however, well beyond the scope of this book to give a complete account of this theory. We have found, for example, Drazin & Johnson's (1989) introduction to soliton theory quite accessible for upper-level undergraduates and beginning graduate students. Readers who are interested in a detailed account of *Inverse Scattering Theory* for soliton equations with one spatial dimension are referred to Ablowitz & Segur (1981) and with higher spatial dimensions to Ablowitz & Clarkson (1991). Another excellent account of soliton theory, with a particular emphasis on the algebraic interpretation of the theory, is given by Newell (1985). The literature on soliton theory and related applications is vast. Additional references can be found in the extensive bibliographies contained in these four books.

6.1 A derivation of the *KdV* equation

The *KdV* equation is a mathematical model for the evolution of weakly-dispersive and weakly-nonlinear waves. It usually arises as a solvability condition associated with an asymptotic expansion for a more general mathematical model similar in spirit, if not in detail, to our derivation of the *CHM* equation in Chapter 5. The *KdV* equation was first derived by Korteweg & de Vries (1985) in a study of long surface gravity waves on a homogenous fluid. This derivation can be found in, for example, LeBlond & Mysak (1978), Newell (1985) or Drazin & Johnson (1989). Here, we derive the *KdV* equation for long small-amplitude internal gravity waves in a continuously stratified fluid of finite depth.

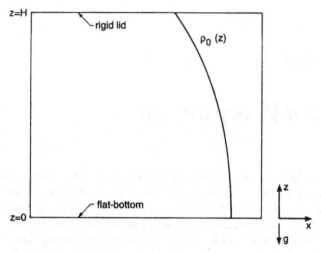

Figure 6.1. Geometry of the configuration used to derive the *KdV* equation.

We begin our analysis with the two dimensional adiabatic equations for an incompressible continuously stratified Boussinesq fluid under the influence of gravity, which can be written in the form (see, e.g., LeBlond & Mysak, 1978 or Kundu, 1990)

$$u_t + uu_x + wu_z = -\frac{1}{\rho_*}p_x, \tag{6.1}$$

$$w_t + uw_x + ww_z = -\frac{1}{\rho_*}p_z - \frac{\rho g}{\rho_*}, \tag{6.2}$$

$$\rho_t + u\rho_x + w\rho_z = 0, \tag{6.3}$$

$$u_x + w_z = 0, \tag{6.4}$$

where u and w correspond to the horizontal and vertical velocities, respectively, x and z correspond to the horizontal and vertical coordinates, respectively (see Fig. 6.1), ρ is the variable density, p is the pressure, g is the gravitational acceleration and ρ_* is a constant density associated with the Boussinesq approximation. The Boussinesq approximation is an approximation relating to holding the density constant *except* in the buoyancy terms in the equations of motion; see LeBlond & Mysak (1978) or Kundu (1990). Equations (6.1) and (6.2) are the horizontal and vertical momentum equations, respectively. Equation (6.3) expresses the fact that the fluid is incompressible. Equation (6.4) is the statement that mass is conserved in the fluid.

In the absence of any motion, i.e.,

$$u = w \equiv 0, \tag{6.5}$$

the fluid is assumed to be in hydrostatic balance with respect to the vertically-stratified background density and pressure fields given by $\rho_o(z)$ and $p_o(z)$, respectively, i.e.,

$$\frac{dp_o(z)}{dz} = -g\rho_o(z), \tag{6.6}$$

where it is assumed that the fluid is stably stratified, i.e.,

$$\frac{d\rho_o(z)}{dz} < 0. \tag{6.7}$$

The reader may verify that (6.5) and (6.6) is an exact nonlinear solution to (6.1) through to (6.4). The constant density ρ_* in the Boussinesq approximation may be chosen as

$$\rho_* = \frac{1}{H} \int_0^H \rho_o(\xi)\, d\xi,$$

where H is the depth of the fluid; see Fig. 6.1.

On account of (6.4), it follows that there exists a stream function $\psi = \psi(x, z, t)$ satisfying

$$u = -\frac{\partial \psi}{\partial z}, \tag{6.8}$$

$$w = \frac{\partial \psi}{\partial x}. \tag{6.9}$$

If the vorticity equation

$$\partial_x(6.2) - \partial_z(6.1),$$

is formed, with the velocity components replaced with the stream function, one obtains

$$\Delta \psi_t + \partial(\psi, \Delta \psi) = -\frac{g}{\rho_*} \rho_x, \tag{6.10}$$

for the vorticity equation, and (6.3) can be written in the form

$$\rho_t + \partial\left(\psi, \rho\right) = 0, \tag{6.11}$$

where $\partial\left(A, B\right) \equiv A_x B_z - A_z B_x$. The pair of partial differential equations (6.10) and (6.11) will determine the evolution of the stream function and density fields and the corresponding velocity field is then determined from (6.8) and (6.9). The pressure will obtained from the Poisson problem obtained from taking the divergence of the momentum equations, i.e.,

$$\partial_x\left(6.1\right) + \partial_z\left(6.2\right).$$

We assume that the fluid is confined between two rigid boundaries at $z = 0$ and $z = H$, respectively. Since the fluid is inviscid, the appropriate boundary condition is $w = 0$ on $z = 0$ and $z = H$, respectively, which implies, on account of (6.9), that $\psi = \psi_o\left(t\right)$ and $\psi_H\left(t\right)$ on $z = 0$ and $z = H$, respectively. We will choose $\psi_o\left(t\right) = \psi_H\left(t\right) \equiv 0$. This may be interpreted as specifying that there is no net mass flux in the x-direction. We will specify boundary and/or periodicity conditions with respect to x as we need them.

To proceed further it is advantageous to nondimensionalize the variables. Let us introduce the following nondimensional primed quantities:

$$x = Lx', \ z = Hz', \ t = \frac{L}{NH}t', \tag{6.12}$$

$$\psi\left(x, z, t\right) = \varepsilon N H^2 \psi'\left(x', z', t'\right), \tag{6.13}$$

$$\rho\left(x, z, t\right) = \rho_o\left(z\right) + \varepsilon \rho_* H N^2 g^{-1} \rho'\left(x', z', t'\right), \tag{6.14}$$

where L is an appropriate horizontal length scale, N is an appropriate scaling for the Brunt-Väisälä frequency

$$\left(-g\rho_*^{-1}\frac{d\rho_o}{dz}\right)^{\frac{1}{2}},$$

and ε is the squared aspect ratio given by

$$\varepsilon \equiv \left(\frac{H}{L}\right)^2. \tag{6.15}$$

The Brunt-Väisälä frequency is the frequency with which an (infinitesimally) vertically displaced fluid parcel oscillates about its equilibrium position in a stably stratified fluid. Note that it follows from (6.14) that $\rho'\left(x', z', t'\right)$ can be interpreted as the departure of the total density field from the background hydrostatic density field.

Substitution of (6.12), (6.13) and (6.14) into (6.10) and (6.11) leads to, after a little algebra and dropping the primes on the nondimensional variables,

$$\rho_t - S(z)\psi_x = -\varepsilon\partial(\psi,\rho),\tag{6.16}$$

$$\psi_{zzt} + \rho_x = -\varepsilon\psi_{xxt} - \varepsilon\partial(\psi,\psi_{zz}) - \varepsilon^2\partial(\psi,\psi_{xx}),\tag{6.17}$$

where the stratification function $S(z)$, or nondimensional squared Brunt-Väisälä frequency, is given by

$$S(z) \equiv -\frac{g}{\rho_* N^2}\frac{d\rho_o(z)}{dz} > 0.\tag{6.18}$$

It is understood in (6.16) and (6.17) that the derivatives in the Jacobian terms are with respect to the nondimensional variables x and z.

The asymptotic approximation that the waves are long (relative to the depth of the fluid) and weakly-nonlinear is realized with the assumption that the aspect ratio, i.e., $\sqrt{\varepsilon}$, satisfies $0 < \varepsilon \ll 1$. We shall consider a slowly-varying solution written with respect to the *fast* co-moving coordinate

$$\tilde{x} = x - ct,\tag{6.19}$$

and the *slow* time variable

$$\tilde{t} = \varepsilon t.\tag{6.20}$$

Substitution (6.19) and (6.20) into (6.15) and (6.16) leads to, after dropping the tildes on $\tilde{x}$ and $\tilde{t}$,

$$c\rho_x + S(z)\psi_x = \varepsilon\partial(\psi,\rho) + \varepsilon\rho_t + O(\varepsilon^2),\tag{6.21}$$

$$\rho_x - c\psi_{zzx} = \varepsilon c\psi_{xxx} - \varepsilon\psi_{zzt} - \varepsilon\partial(\psi,\psi_{zz}) + O(\varepsilon^2),\tag{6.22}$$

where it is understood that the derivatives in the Jacobian terms are written with respect to x and z.

Introducing the straightforward asymptotic expansion

$$\rho(x,z,t) \simeq \rho^{(0)}(x,z,t) + \varepsilon\rho^{(1)}(x,z,t) + O(\varepsilon^2),$$

$$\psi(x,z,t) \simeq \psi^{(0)}(x,z,t) + \varepsilon\psi^{(1)}(x,z,t) + O(\varepsilon^2),$$

into (6.21) and (6.22) leads to the $O(1)$ problem

$$\rho^{(0)} = -\frac{S(z)\psi^{(0)}}{c},\tag{6.23}$$

$$\psi_{zz}^{(0)} + \frac{S(z)}{c^2}\psi^{(0)}, \tag{6.24}$$

where we have integrated (6.22) with respect to x and set the integration constant to zero to be consistent with the no net x-direction mass flux assumption.

Equation (6.24) possesses a separated normal-mode solution of the form

$$\psi(x, z, t) = \varphi(x, t)\,\Phi(z).$$

Substitution of this decomposition into (6.24) leads to the two-point boundary value problem

$$\Phi_{zz} + \frac{S(z)}{c^2}\Phi = 0, \tag{6.25}$$

$$\Phi(0) = \Phi(1) = 0. \tag{6.26}$$

This is a Sturm-Liouville eigenvalue problem for which we are guaranteed that c is real and nonzero (see, e.g., Zauderer, 1983). Let us assume that we have chosen a particular eigenfunction-eigenvalue pair (Φ, c).

The $O(\varepsilon)$ problem associated with the asymptotic expansion can be written in the form

$$\rho_x^{(1)} + \frac{S(z)\psi_x^{(1)}}{c} = \frac{\partial\left(\psi^{(0)}, \rho^{(0)}\right) + \rho_t^{(0)}}{c},$$

$$\psi_{zzx}^{(1)} - \frac{\rho_x^{(1)}}{c} = -\psi_{xxx}^{(0)} + \frac{\psi_{zzt}^{(0)} + \partial\left(\psi^{(0)}, \psi_{zz}^{(0)}\right)}{c}.$$

If $\rho^{(1)}$ is eliminated between these two expressions it follows that

$$\psi_{zzx}^{(1)} + \frac{S(z)}{c^2}\psi_x^{(1)} = \frac{\psi_{zzt}^{(0)} + \partial\left(\psi^{(0)}, \psi_{zz}^{(0)}\right)}{c^2}$$

$$\frac{\psi_{zzt}^{(0)} + \partial\left(\psi^{(0)}, \psi_{zz}^{(0)}\right)}{c} - \psi_{xxx}^{(0)}$$

$$= -\left(\frac{2S(z)\varphi_t}{c^3} + \varphi_{xxx}\right)\Phi - \left(\frac{2\varphi\varphi_x S_z}{c^3}\right)\Phi^2. \tag{6.27}$$

If we multiply the left and right sides of (6.27) through by Φ and integrate with respect to z over the interval $z \in (0,1)$, it follows that the integral associated with the left hand side can be integrated by parts in the form

$$\int_0^1 \Phi \left(\psi_{zzx}^{(1)} + \frac{S(z)}{c^2} \psi_x^{(1)} \right) dz =$$

$$\left[\Phi(z) \psi_{zx}^{(1)}(x,z,t) - \Phi_z(z) \psi_x^{(1)}(x,z,t) \right]_{z=0}^{z=1}$$

$$+ \int_0^1 \psi_x^{(1)} \left(\Phi_{zz} + \frac{S(z)}{c^2} \Phi \right) dz = 0,$$

since the boundary terms are identically zero because of the boundary conditions $\psi^{(1)} = \Phi \equiv 0$ on $z = 0$ and 1, respectively, and because (6.25) holds.

Thus, we conclude from the integral associated with the right hand side of (6.27) that

$$\varphi_t + \alpha\varphi\varphi_x + \beta\varphi_{xxx} = 0, \tag{6.28}$$

where

$$\alpha \equiv \frac{\displaystyle\int_0^1 S_z(z) \Phi^3(z) \, dz}{\displaystyle\int_0^1 S(z) \Phi(z) \, dz}, \tag{6.29}$$

$$\beta \equiv \frac{c^3 \displaystyle\int_0^1 \Phi^2(z) \, dz}{2 \displaystyle\int_0^1 S(z) \Phi^2(z) \, dz}. \tag{6.30}$$

Equation (6.28) is a *Korteweg-de Vries* or *KdV* equation. We may write the *KdV* equation in its standard form if we introduce the transformation

$$\varphi(x,t) = \beta\alpha^{-1} q\left(x, \hat{t}\right), \tag{6.31}$$

$$t = \beta^{-1}\hat{t}. \tag{6.32}$$

Substitution of (6.31) and (6.32) into (6.28) leads to

$$q_t + qq_x + q_{xxx} = 0, \tag{6.33}$$

where we have dropped the caret on the new time variable. Equation (6.33) will be the form of the *KdV* equation that we will henceforth restrict our attention to.

If the fluid has constant stratification, i.e., S is constant, then it follows from (6.29) that $\alpha = 0$. In this case the evolution is not governed by the *KdV* equation (6.33), but rather by a higher order modified Korteweg-de Vries (*mKdV*) equation in which the quadratic nonlinear term in (6.33) is replaced with a cubic or higher order nonlinearity (see, e.g., Gear & Grimshaw, 1983).

Having derived the *KdV* equation it is convenient to specify the boundary conditions with respect to x that will be considered in this chapter. There are two configurations of interest. One domain and accompanying boundary conditions that we shall work with is the infinite interval

$$\Omega_\infty \equiv \{x \mid -\infty < x < \infty\}, \tag{6.34}$$

with the assumption that $q(x,t)$ and all its derivatives are continuous square-integrable functions on Ω_∞.

The second domain and accompanying boundary conditions that we will work with is the finite periodic domain

$$\Omega_o \equiv \{x \mid -x_o < x < x_o\}, \tag{6.35}$$

with the assumption that $q(x,t)$ and all its derivatives are continuous functions on Ω_o and are periodic at $\pm x_o$. As a notational convenience, we will denote the spatial domain as simply Ω when the argument being presented holds for either Ω_∞ or Ω_o. The reader who is interested in a general account of the mathematical theory of the Cauchy problem associated with the *KdV* equation is referred to Bona & Smith (1975).

6.2   Hamiltonian structure

Hamiltonian formulation

The Hamiltonian structure for the *KdV* equation was first described by Gardner (1971). Subsequently, Zakharov & Faddeev (1971) showed that the *KdV* equation was a completely integrable Hamiltonian system. Here, again, we remark that there only exists a variational principle for the *KdV* equation in terms of Eulerian variables if a Clebsch variable transformation is introduced (see, e.g., Atherton & Homsy, 1975; Grimshaw, 1979 and the exercises).

Theorem 6.1 *The KdV equation (6.33) is Hamiltonian for the choice of*

$$H(q) = \frac{1}{2} \int_\Omega q_x^2 - \frac{q^3}{3} \, dx, \qquad (6.36)$$

$$J = \frac{\partial}{\partial x}. \qquad (6.37)$$

Proof. We will present a complete proof of this theorem. The first thing we wish to show is that $H(q)$ is an invariant of the motion. It follows that

$$\frac{dH}{dt} = \int_\Omega q_x q_{xt} - \frac{q^2 q_t}{2} \, dx$$

$$= -\int_\Omega \left(q_{xx} + \frac{q^2}{2} \right) q_t \, dx, \qquad (6.38)$$

where we integrated by parts once exploiting the periodicity conditions if $\Omega = \Omega_o$ or the fact that it is assumed that $|q| \to 0$ smoothly as $|x| \to \infty$ if $\Omega = \Omega_\infty$. Substituting (6.33) into (6.38) leads to

$$\frac{dH}{dt} = \int_\Omega q q_x q_{xx} + q_{xx} q_{xxx} + \frac{q^3 q_x + q^2 q_{xxx}}{2} \, dx. \qquad (6.39)$$

Observing that

$$q q_x q_{xx} + \frac{q^2 q_{xxx}}{2} = \frac{1}{2} \left(q^2 q_{xx} \right)_x,$$

$$q_{xx} q_{xxx} + \frac{q^3 q_x}{2} = \frac{1}{2} \left(q_{xx}^2 + \frac{q^4}{4} \right)_x,$$

allows (6.39) to be re-written in the form

$$\frac{dH}{dt} = \frac{1}{2} \int_\Omega \left(q^2 q_{xx} + q_{xx}^2 + \frac{q^4}{4} \right)_x dx = 0,$$

which establishes that $H(q)$ is an invariant of the motion.

We next show that the KdV equation is reproduced by the expression

$$q_t = J \frac{\delta H}{\delta q}.$$

The variational derivative $\frac{\delta H}{\delta q}$ is obtained from the first variation δH given by

$$\delta H = \int_\Omega q_x \delta q_x - \frac{q^2 \delta q}{2} \, dx$$

$$= -\int_\Omega \left(q_{xx} + \frac{q^2}{2} \right) \delta q \, dx, \tag{6.40}$$

after an integration by parts. Thus, we conclude that

$$\frac{\delta H}{\delta q} = -\left(q_{xx} + \frac{q^2}{2} \right),$$

which implies that

$$J \frac{\delta H}{\delta q} = -\left(q_{xx} + \frac{q^2}{2} \right)_x$$

$$= -q q_x - q_{xxx} = q_t.$$

We next establish that the bracket

$$[F, G] \equiv \left\langle \frac{\delta F}{\delta q}, J \frac{\delta G}{\delta q} \right\rangle$$

$$= \int_\Omega \frac{\delta F}{\delta q} \frac{\partial}{\partial x} \left(\frac{\delta G}{\delta q} \right) \, dx, \tag{6.41}$$

where F and G are arbitrary functionals of q, satisfies the five required algebraic properties *H1, H2, H3, H4* and *H5* given in Section 3.2 for a Hamiltonian system. It is understood, of course, that F and G satisfy the appropriate boundary or periodicity conditions associated with Ω. Such a functional is referred to as an *allowable functional*.

Self-commutation (property *H1*) follows immediately since

$$[F, F] = \int_\Omega \frac{\delta F}{\delta q} \frac{\partial}{\partial x} \left(\frac{\delta F}{\delta q} \right) \, dx$$

$$= \frac{1}{2} \int_\Omega \frac{\partial}{\partial x} \left[\left(\frac{\delta F}{\delta q} \right)^2 \right] \, dx = 0,$$

where we have exploited the fact that

$$\left| \frac{\delta F}{\delta q} \right| \to 0 \text{ as } |x| \to \infty,$$

if $\Omega = \Omega_\infty$ or the periodicity of $\frac{\delta F}{\delta q}$ at $x = \pm x_o$ if $\Omega = \Omega_o$.

Skew-symmetry (property *H2*) is a consequence of a simple integration by parts, i.e.,

$$[F, G] = \int_\Omega \frac{\delta F}{\delta q} \frac{\partial}{\partial x} \left(\frac{\delta G}{\delta q} \right) dx$$

$$= - \int_\Omega \frac{\delta G}{\delta q} \frac{\partial}{\partial x} \left(\frac{\delta F}{\delta q} \right) dx = 0,$$

where the boundary integral is zero for either $\Omega = \Omega_\infty$ or Ω_o.

The distributive property (*H3*) follows from the linearity of the variational derivative and the integral, i.e.,

$$[\alpha F + \beta G, Q] = \int_\Omega \frac{\delta (\alpha F + \beta G)}{\delta q} \frac{\partial}{\partial x} \left(\frac{\delta Q}{\delta q} \right) dx$$

$$= \alpha \int_\Omega \frac{\delta F}{\delta q} \frac{\partial}{\partial x} \left(\frac{\delta Q}{\delta q} \right) dx$$

$$+ \beta \int_\Omega \frac{\delta G}{\delta q} \frac{\partial}{\partial x} \left(\frac{\delta Q}{\delta q} \right) dx$$

$$= \alpha [F, Q] + \beta [G, Q],$$

where α and β are arbitrary real constants and Q is an arbitrary allowable functional of q.

The associative property (*H4*) follows in a similar manner, i.e.,

$$[FG, Q] = \int_\Omega \frac{\delta (FG)}{\delta q} \frac{\partial}{\partial x} \left(\frac{\delta Q}{\delta q} \right) dx$$

$$= \int_\Omega \left(\frac{\delta F}{\delta q} G + F \frac{\delta G}{\delta q} \right) \frac{\partial}{\partial x} \left(\frac{\delta Q}{\delta q} \right) dx$$

$$= \left[\int_\Omega \frac{\delta F}{\delta q} \frac{\partial}{\partial x} \left(\frac{\delta Q}{\delta q} \right) \, dx \right] G$$

$$+ F \left[\int_\Omega \frac{\delta G}{\delta q} \frac{\partial}{\partial x} \left(\frac{\delta Q}{\delta q} \right) \, dx \right]$$

$$= [F, Q] \, G + F \, [G, Q] \,,$$

where the fact that F and G, as functionals, are functions of time only and can therefore be pulled outside the integral.

The final property $(H5)$ that must be shown is the Jacobi identity, i.e.,

$$[[G, Q], F] + [[Q, F], G] + [[F, G], Q] = 0.$$

To establish this identity we proceed directly,

$$[[G, Q], F] + [[Q, F], G] + [[F, G], Q]$$

$$= \int_\Omega \frac{\delta}{\delta q} [G, Q] \frac{\partial}{\partial x} \left(\frac{\delta F}{\delta q} \right) \, dx$$

$$+ \int_\Omega \frac{\delta}{\delta q} [Q, F] \frac{\partial}{\partial x} \left(\frac{\delta G}{\delta q} \right) \, dx$$

$$+ \int_\Omega \frac{\delta}{\delta q} [F, G] \frac{\partial}{\partial x} \left(\frac{\delta Q}{\delta q} \right) \, dx$$

$$= \int_\Omega \frac{\delta}{\delta q} \left\{ \left\langle \frac{\delta G}{\delta q}, \frac{\partial}{\partial x} \left(\frac{\delta Q}{\delta q} \right) \right\rangle \right\} \frac{\partial}{\partial x} \left(\frac{\delta F}{\delta q} \right) \, dx$$

$$+ \int_\Omega \frac{\delta}{\delta q} \left\{ \left\langle \frac{\delta Q}{\delta q}, \frac{\partial}{\partial x} \left(\frac{\delta F}{\delta q} \right) \right\rangle \right\} \frac{\partial}{\partial x} \left(\frac{\delta G}{\delta q} \right) \, dx$$

$$+ \int_\Omega \frac{\delta}{\delta q} \left\{ \left\langle \frac{\delta F}{\delta q}, \frac{\partial}{\partial x} \left(\frac{\delta G}{\delta q} \right) \right\rangle \right\} \frac{\partial}{\partial x} \left(\frac{\delta Q}{\delta q} \right) \, dx$$

$$= \int_\Omega \frac{\delta}{\delta q} \{\langle G', Q'_x \rangle\} F'_x \, dx$$

$$+ \int_\Omega \frac{\delta}{\delta q} \{\langle Q', F'_x \rangle\} G'_x \, dx$$

$$+ \int_\Omega \frac{\delta}{\delta q} \{\langle F', G'_x \rangle\} Q'_x \, dx, \tag{6.42}$$

where we have introduced

$$(*)' \equiv \frac{\delta(*)}{\delta q} \text{ and } (*)'_x \equiv \frac{\partial}{\partial x} \left\{ \frac{\delta(*)}{\delta q} \right\}.$$

To simplify (6.42) further we must compute the variational derivatives

$$\frac{\delta}{\delta q} \langle G', Q'_x \rangle, \quad \frac{\delta}{\delta q} \langle Q', F'_x \rangle \text{ and } \frac{\delta}{\delta q} \langle F', G'_x \rangle.$$

We will explicitly only compute the first of these and the others will follow in a similar manner. We begin with computing the first variation of $\langle G', Q'_x \rangle$ as follows

$$\delta \langle G', Q'_x \rangle = \delta \int_\Omega G' Q'_x \, dx$$

$$= \int_\Omega G'' Q'_x \delta q + G' (Q'' \delta q)_x \, dx$$

$$= \int_\Omega \{ G'' Q'_x - G'_x Q'' \} \delta q \, dx,$$

where, for convenience, we write

$$\delta (*)' = (*)'' \delta q \text{ with } (*)'' \equiv \frac{\delta^2(*)}{\delta q^2},$$

and we have integrated by parts once. We therefore conclude that

$$\frac{\delta}{\delta q} \langle G', Q'_x \rangle = G'' Q'_x - G'_x Q'', \tag{6.43}$$

and similarly that

$$\frac{\delta}{\delta q} \langle Q', F'_x \rangle = Q'' F'_x - Q'_x F'', \tag{6.44}$$

$$\frac{\delta}{\delta q} \langle F', G'_x \rangle = F'' G'_x - F'_x G''. \tag{6.45}$$

Substitution of (6.43), (6.44) and (6.45) into (6.42) implies that

$$[[G, Q], F] + [[Q, F], G] + [[F, G], Q]$$

$$= \int_{\Omega} \{G'' Q'_x - G'_x Q''\} F'_x \, dx$$

$$+ \int_{\Omega} \{Q'' F'_x - Q'_x F''\} G'_x \, dx$$

$$+ \int_{\Omega} \{F'' G'_x - F'_x G''\} Q'_x \, dx = 0,$$

since the integrands sum to identically zero. This completes the proof. ∎

Casimir and Momentum invariants

The KdV equation has an infinite number of conserved functionals associated with it (Miura *et al.*, 1968). However, only a few of these conserved functionals have any obvious direct physical interpretation such conservation of energy, mass, momentum, and so on. For the purposes of our discussion it is sufficient to establish only the conserved functional associated with the invariance of the Hamiltonian structure to arbitrary translations in x, that is, momentum conservation. In the case where one is working with the periodic domain Ω_o it is understood that the x-translations are modulo $2x_o$.

As a consequence of Noether's Theorem 3.5, the time-invariant functional, denoted $M(q)$, associated with the invariance of the Hamiltonian structure to arbitrary translations in x satisfies

$$J \frac{\delta M}{\delta q} = -q_x,$$

that is,

$$\frac{\partial}{\partial x}\left(\frac{\delta M}{\delta q}\right) = -q_x. \tag{6.46}$$

It follows that

$$\frac{\delta M}{\delta q} = -q,$$

to within a constant, and thus that

$$M(q) = -\frac{1}{2}\int_\Omega q^2 dx. \tag{6.47}$$

Even though Noether's Theorem guarantees that $M(q)$ is invariant with respect to time it is useful to verify this directly. This is left as an exercise for the reader. Also, it is left as an exercise for the reader to check that the conserved functional, as obtained by applying Noether's Theorem, associated with the invariance of the Hamiltonian structure with respect to arbitrary time translations is $-H(q)$.

The Casimir functionals, denoted $C(q)$, satisfy

$$J\frac{\delta C}{\delta q} = 0,$$

that is,

$$\frac{\partial}{\partial x}\left(\frac{\delta C}{\delta q}\right) = 0, \tag{6.48}$$

which implies that $\frac{\delta C}{\delta q}$ is constant, so that

$$C(q) = \int_\Omega q\, dx. \tag{6.49}$$

Alternate Hamiltonian formulation

The *KdV* equation (6.33) possesses another Hamiltonian structure. This structure is given by

$$H = -\frac{1}{2}\int_\Omega q^2\, dx, \tag{6.50}$$

$$J = \frac{\partial^3}{\partial x^3} + \frac{1}{3}\left(2q\frac{\partial}{\partial x} + q_x\right). \tag{6.51}$$

Note that the Hamiltonian in this formulation is the momentum invariant associated with the previous Hamiltonian formulation. It is left as an exercise for the reader to verify that (6.50) and (6.51) does indeed constitute a Hamiltonian formulation of the *KdV* equation.

The dual Hamiltonian structure of the *KdV* equation is characteristic of all known integrable infinite dimensional dynamical systems (see, e.g., Newell, 1985). It is conjectured, but at this time still unproven, that a dual Hamiltonian structure is a necessary condition for an infinite dimensional dynamical system to be integrable in the sense of admitting an *Inverse Scattering Transform*.

6.3   Periodic and soliton solutions

In this section we will present a brief derivation of the periodic and soliton solutions of the *KdV* equation (6.33). Readers will find a more complete presentation in, for example, Ablowitz & Segur (1981) or Drazin & Johnson (1989).

General periodic solution

We begin by assuming a steadily-travelling solution to (6.33) of the form

$$q\left(x,t\right) = -6\Phi\left(\xi\right) \text{ with } \xi \equiv x - ct, \tag{6.52}$$

where c is the translation velocity (note that this translation velocity is *not* the translation velocity introduced earlier in our original derivation of the *KdV* equation in Section 6.1). Substitution of (6.52) into (6.33) leads to the ordinary differential equation

$$\Phi^{'''} = 6\Phi\Phi^{'} + c\Phi^{'},$$

where

$$\Phi^{'}\left(\xi\right) \equiv \frac{d\Phi\left(\xi\right)}{d\xi}.$$

This equation can be integrated once with respect to ξ to yield

$$\Phi^{''} = 3\Phi^2 + c\Phi + A, \tag{6.53}$$

where A is a free constant of integration.

Multiplying (6.53) through by $\Phi^{'}$ and integrating with respect to ξ again yields

$$\left(\Phi^{'}\right)^2 = 2\Pi\left(\Phi\right), \tag{6.54}$$

where $\Pi\left(\Phi\right)$ is the cubic polynomial given by

$$\Pi\left(\Phi\right)=\Phi^3+\frac{c\Phi^2}{2}+A\Phi+B, \tag{6.55}$$

where B is another free constant of integration. It follows from (6.54) that

$$\xi-\xi_0=\int_{\Phi_0}^{\Phi}\frac{d\eta}{\sqrt{2\Pi\left(\eta\right)}}, \tag{6.56}$$

where ξ_0 and Φ_0 are integration constants satisfying $\Phi\left(\xi_0\right)=\Phi_0$.

It can be shown (see, for example, the geometrical argument presented in Drazin & Johnson, 1989) that real bounded solutions for $\Phi\left(\xi\right)$ correspond to values of the (A,B,c) parameters for which $\Pi\left(\Phi\right)$ has three real roots, with at least one different than the other two. Let us denote these real roots as Φ_1, Φ_2 and Φ_3, respectively, with the ordering

$$\Phi_3<\Phi_2<\Phi_1,$$

then (6.56) may be re-written in the form

$$\xi-\xi_0=\int_{\Phi_3}^{\Phi}\frac{d\eta}{[2\left(\eta-\Phi_3\right)\left(\eta-\Phi_2\right)\left(\eta-\Phi_1\right)]^{\frac{1}{2}}}, \tag{6.57}$$

where we have chosen $\Phi_0=\Phi_3$. If we introduce the transformation

$$\eta=\Phi_3+\left(\Phi_2-\Phi_3\right)\sin^2\left(\zeta\right), \tag{6.58}$$

into (6.57), the integral may be written in the form, after a little algebra,

$$\left(\frac{\Phi_1-\Phi_3}{2}\right)^{\frac{1}{2}}\left(\xi-\xi_0\right)=\int_0^{\theta}\frac{d\zeta}{\sqrt{1-m\sin^2\left(\zeta\right)}}, \tag{6.59}$$

where the parameter m is given by

$$0<m\equiv\frac{\Phi_2-\Phi_3}{\Phi_1-\Phi_3}<1, \tag{6.60}$$

and where the upper limit of integration in (6.59), given by θ, satisfies

$$\Phi=\Phi_3+\left(\Phi_2-\Phi_3\right)\sin^2\left(\theta\right)$$

$$=\Phi_2-\left(\Phi_2-\Phi_3\right)\cos^2\left(\theta\right). \tag{6.61}$$

The integral on the right-hand-side of (6.59) is an elliptic integral of the first kind (see, for example, Abramowitz & Stegun, 1972). It is convenient to rewrite (6.59) in terms of the Jacobian elliptic *cnoidal* $cn(\psi \mid m)$ function, defined by the pair of equations

$$cn(\psi \mid m) = \cos(\theta), \tag{6.62}$$

$$\psi = \int_0^\theta \frac{d\zeta}{\sqrt{1 - m\sin^2(\zeta)}}. \tag{6.63}$$

It follows from (6.59), (6.61), (6.62) and (6.63) that we may write $\Phi(\xi)$ in the form

$$\Phi(\xi) = \Phi_2 - (\Phi_2 - \Phi_3)\, cn^2 \left[\sqrt{\frac{(\Phi_1 - \Phi_3)}{2}}(\xi - \xi_0) \mid m \right]. \tag{6.64}$$

The steadily travelling solution (6.64) corresponds to a periodic function which oscillates between the values Φ_3 and Φ_2. Since $\cos^2(\theta)$ is periodic with period π, it follows from (6.59) that the *period* of the oscillation with respect to the ξ-variable, denoted λ_ξ, will be given by

$$\left(\frac{\Phi_1 - \Phi_3}{2}\right)^{\frac{1}{2}} \lambda_\xi = \int_0^\pi \frac{d\zeta}{\sqrt{1 - m\sin^2(\zeta)}}$$

$$= 2\int_0^{\frac{\pi}{2}} \frac{d\zeta}{\sqrt{1 - m\sin^2(\zeta)}} = 2K(m),$$

where $K(m)$ is the complete elliptic integral of the first kind defined by

$$K(m) = \int_0^{\frac{\pi}{2}} \frac{d\zeta}{\sqrt{1 - m\sin^2(\zeta)}},$$

and where we have exploited the fact that $\sin^2(\zeta)$ is an even function about $\zeta = \frac{\pi}{2}$. It therefore follows that

$$\lambda_\xi = 2K(m) \left(\frac{2}{\Phi_1 - \Phi_3}\right)^{\frac{1}{2}}. \tag{6.65}$$

Note that this expression explicitly requires that $\Phi_1 \neq \Phi_3$, that is, $m \neq 1$. In terms of the x and t variables in the *KdV* equation (6.33), the solution described by (6.52) and (6.64) has a *wavelength* and *temporal period,* denoted by λ_x and P_t, respectively, given by

$$\lambda_x = \lambda_\xi, \tag{6.66}$$

$$P_t = \lambda_\xi/c, \tag{6.67}$$

respectively.

Soliton solution

The soliton solution we construct here for the *KdV* equation (6.33) assumes that $q(x,t)$ and all its derivative are square-integrable functions for all time over $x \in (-\infty, \infty)$. Thus, if we assume a steadily-travelling solution of the form (6.52), it follows that $A = B = 0$ in (6.53) and (6.55) since $q(x,t)$ and all its derivatives necessarily vanish as $|x| \to \infty$. Consequently (6.54) can be written in the form

$$\frac{d\Phi}{d\xi} = \pm\Phi(2\Phi + c)^{\frac{1}{2}}. \tag{6.68}$$

This ordinary differential equation can be easily integrated after we introduce the transformation

$$\Phi(\xi) = \frac{\Upsilon^2(\xi) - c}{2}, \tag{6.69}$$

which leads to the differential equation

$$\frac{d\Upsilon}{d\xi} = \pm\frac{\Upsilon^2 - c}{2}, \tag{6.70}$$

and thus

$$\int \frac{d\Upsilon}{\Upsilon^2 - c} = \pm\frac{\xi - \xi_0}{2}, \tag{6.71}$$

where ξ_0 is a constant of integration.

The bounded solution for Υ corresponds to assuming

$$0 < \Upsilon^2 < c,$$

which implies that

$$\int \frac{d\Upsilon}{\Upsilon^2 - c} = -\frac{1}{\sqrt{c}} \tanh^{-1}\left(\frac{\Upsilon}{\sqrt{c}}\right), \tag{6.72}$$

which together with (6.71) gives

$$\Upsilon = \mp\sqrt{c}\tanh\left[\frac{\sqrt{c}\,(\xi - \xi_0)}{2}\right].\tag{6.73}$$

Substitution of (6.73) into (6.69) yields

$$\Phi\left(\xi\right) = -\frac{c}{2}\operatorname{sech}^2\left[\frac{\sqrt{c}\,(\xi - \xi_0)}{2}\right].\tag{6.74}$$

Substituting (6.74) into (6.52) gives

$$q(x,t) = q_s\left(x - ct\right) \equiv 3c\operatorname{sech}^2\left[\frac{\sqrt{c}\,(x - ct - \xi_0)}{2}\right],\tag{6.75}$$

which is the soliton solution to the *KdV* equation (6.33) which decays to zero at infinity. Note that $c > 0$ for this solution so that the soliton moves in the direction of increasing x.

There is a slight generalization of this solution which does not approach zero at infinity but rather smoothly asymptotes to a nonzero constant, i.e., $q \to q_\infty \neq 0$, as $|x| \to \infty$. Since the *KdV* equation (6.33) is invariant under the Galilean transformation

$$q\left(x,t\right) = q_\infty + \widetilde{q}\left(\widetilde{x},\widetilde{t}\right),$$

where $\widetilde{x} = x - q_\infty t$ and $\widetilde{t} = t$ for any value of q_∞ (this demonstration is left as an exercise for the reader), it follows that this solution can be written in the form

$$q\left(x,t\right) = q_\infty - 6\Phi\left(\xi - q_\infty t\right),\tag{6.76}$$

where $\Phi\left(*\right)$ is given by (6.74). This solution is often described as a *soliton on a background*.

6.4   Variational principles

In this subsection we present variational principles for the soliton and general periodic wave solution to the *KdV* equation (6.33), respectively. We begin with the soliton solution (6.75).

Soliton variational principle

Substituting a solution of the form

$$q = q_s\left(x - ct\right),$$

into (6.33), implies

$$q_{s_{xxx}} + (q_s - c) q_{s_x} = 0, \tag{6.77}$$

which can be integrated with respect to x to give the nonlinear second order ordinary differential equation

$$q_{s_{xx}} = cq_s - \frac{q_s^2}{2}, \tag{6.78}$$

where we have exploited the boundary condition $|q_s| \to 0$ as $|x| \to \infty$. Equation (6.78) is just (6.53) written in terms of q_s (see (6.52)) with $A = 0$.

Consider the constrained Hamiltonian, denoted $\mathcal{H}(q)$, given by

$$\mathcal{H}(q) = H(q) - cM(q)$$

$$= \frac{1}{2} \int_{\Omega_\infty} q_x^2 + cq^2 - \frac{q^3}{3} \, dx,$$

where $H(q)$ is the Hamiltonian (6.36) and $M(q)$ is the linear x-momentum functional (6.47) and where we have chosen Ω to be the unbounded domain Ω_∞. Clearly, $\mathcal{H}(q)$ is an invariant of the KdV equation since both $H(q)$ and $M(q)$ are individually invariant, i.e.,

$$\frac{\partial \mathcal{H}(q)}{\partial t} = \frac{\partial H(q)}{\partial t} + \frac{\partial M(q)}{\partial t} = 0.$$

The first variation of $\mathcal{H}(q)$, denoted $\delta\mathcal{H}(q)$, is given by

$$\delta\mathcal{H}(q) = \int_{\Omega_\infty} q_x \delta q_x + \left(cq - \frac{q^2}{2} \right) \delta q \, dx$$

$$= \int_{\Omega_\infty} \left(cq - \frac{q^2}{2} - q_{xx} \right) \delta q \, dx, \tag{6.79}$$

where we have integrated by parts exploiting the boundary condition that q and δq smoothly vanish at infinity. It follows from (6.78) that

$$\delta\mathcal{H}(q_s) = \int_{\Omega_\infty} \left(cq_s - \frac{q_s^2}{2} - q_{s_{xx}} \right) \delta q \, dx = 0.$$

We have therefore established the variational principle

Theorem 6.2 *The soliton solution given by (6.75), or equivalently (6.78), of the KdV equation (6.33) satisfies the first-order conditions for an extremum of the constrained Hamiltonian functional*

$$\mathcal{H}(q) = \frac{1}{2} \int_{\Omega_\infty} q_x^2 + cq^2 - \frac{q^3}{3} \, dx. \tag{6.80}$$

Periodic wave variational principle

The variational principle associated with a periodic solution to the *KdV* equation differs from that just described for the soliton solution in that, as we now show, it requires, in general, an additional Casimir functional as a constraint. If (6.77) is integrated once with respect to x, one obtains

$$q_{s_{xx}} = cq_s - \frac{q_s^2}{2} + A, \qquad (6.81)$$

where A is free constant of integration which, in general, is not equal to zero. Equation (6.81) is identical to (6.53) except that (6.81) is written with respect to q_s.

Consider the constrained Hamiltonian given by

$$\widehat{\mathcal{H}}(q) = H(q) - cM(q) + AC(q)$$

$$= \frac{1}{2} \int_{\Omega_o} q_x^2 + cq^2 - \frac{q^3}{3} + Aq \, dx,$$

where $H(q)$ and $C(q)$ are as given in (6.81), and where $C(q)$ is the Casimir functional (6.49). Since we are concerned here with the periodic wave solution, we have chosen Ω to be the periodic domain Ω_o. It is understood that the length of the periodic domain Ω_o, i.e., $2x_o$, is an integer multiple of the periodicity of the periodic solution q_s. Clearly, $\widehat{\mathcal{H}}(q)$ is an invariant of the *KdV* equation since each of $H(q)$, $M(q)$ and $C(q)$ is individually invariant.

The first variation of $\widehat{\mathcal{H}}(q)$, denoted $\delta\widehat{\mathcal{H}}(q)$, is given by

$$\delta\widehat{\mathcal{H}}(q) = \int_{\Omega_o} q_x \delta q_x + \left(cq - \frac{q^2}{2} + A\right)\delta q \, dx$$

$$= \int_{\Omega_o} \left(A + cq - \frac{q^2}{2} - q_{xx}\right)\delta q \, dx, \qquad (6.82)$$

where we have integrated by parts exploiting the boundary condition that q and δq are smoothly periodic at $\pm x_o$. It follows from (6.81) and (6.82) that $\delta\widehat{\mathcal{H}}(q_s) = 0$.

We have therefore established the variational principle

Theorem 6.3 *The periodic solution given by (6.64), or equivalently (6.81), of the KdV equation (6.33) satisfies the first-order conditions for an extremum of the constrained Hamiltonian*

$$\widehat{\mathcal{H}}(q) = \frac{1}{2}\int_{\Omega_o} q_x^2 + cq^2 - \frac{q^3}{3} + Aq \, dx. \qquad (6.83)$$

6.5 Linear stability

Before proceeding to develop the nonlinear stability theory it is instructive to first look at the linear stability problem. Our discussion will be focussed on the soliton solution. The stability theory related to the more general periodic solutions of the KdV equation is similar. Details for these periodic solutions can be found in Benjamin (1974) and Drazin (1977).

It is convenient to develop the stability theory for the KdV soliton in a reference moving with the unperturbed soliton. Thus, introducing the co-moving spatial variable

$$\widetilde{x} \equiv x - ct - \xi_0,$$

into (6.33) leads to

$$q_t + (q - c)\,q_x + q_{xxx} = 0, \tag{6.84}$$

where we have deleted the tilde from $\widetilde{x}$. The KdV soliton corresponds, therefore, to a steady solution of (6.84). In the remainder of this section and throughout the next section we shall work in terms of $\widetilde{x}$ (with the tilde deleted) unless otherwise specified.

The introduction of the above co-moving reference frame modifies the Hamiltonian formulation. The Hamiltonian formulation for (6.84) is no longer described by Theorem 6.1. It is, however, straightforward to verify the Hamiltonian functional associated with (6.84) is simply $\mathcal{H}(q)$ given by (6.80) with $J = -\partial_x$. Since $\mathcal{H}(q)$ is invariant to arbitrary translations in x, we may consider the integration variable in $\mathcal{H}(q)$ to be the co-moving variable $\widetilde{x}$ (with the tilde deleted for convenience).

The linear stability equation associated with the KdV soliton is obtained by substituting

$$q = q_s(x) + \delta q(x, t), \tag{6.85}$$

into (6.84), where $q_s(x)$ is given by (6.75), i.e.,

$$q_s(x) = 3c\,\mathrm{sech}^2\left[\frac{\sqrt{c}\,x}{2}\right],$$

and neglecting all nonlinear terms in δq, which yields,

$$\delta q_t + (q_s - c)\,\delta q_x + q_{s_x}\,\delta q + \delta q_{xxx} = 0. \tag{6.86}$$

Note that the subscript s denotes the soliton solution but that subscripts with respect to x and t denote partial differentiation. The appropriate boundary conditions on $\delta q(x, t)$ are that it and all its partial derivatives vanish at infinity sufficiently rapidly so as to be square integrable. We remind the reader that the coefficients involving q_s are appear as time independent functions in (6.86).

Since $q_s(x)$ satisfies the first order necessary condition for an extremal of the conserved functional $\mathcal{H}(q)$, i.e., $\delta\mathcal{H}(q_s) = 0$, linear stability can be established if

$$\delta^2\mathcal{H}(q_s) = \int_{\Omega_\infty} (\delta q_x)^2 + (c - q_s)(\delta q)^2 \; dx$$

$$= \int_{\Omega_\infty} \delta q\,(-\partial_{xx} + c - q_s)\,\delta q\; dx, \tag{6.87}$$

is definite for all perturbations $\delta q(x, t)$.

We can immediately see a problem in developing the stability theory for the *KdV* soliton. It follows from (6.75) that there will always exist values of x for which q_s takes on *any* value in the interval $(0, 3c]$. Thus the coefficient $(c - q_s)$ in the integrand in (6.87) takes on positive and negative values in the integration domain. This coefficient is therefore not itself definite and we cannot claim, on the basis of inspection alone, that $\delta^2\mathcal{H}(q_s)$ is definite. One of Benjamin's (1972) principal contributions was to systematically exploit the spectral properties of the operator

$$\mathcal{L} \equiv -\partial_{xx} + c - q_s(x), \tag{6.88}$$

to show that it was possible to bound $\delta^2\mathcal{H}(q_s)$ away from zero in terms of suitable norms on the perturbation field.

It is left as an exercise for the reader to verify that

$$\delta^2\widehat{\mathcal{H}}(q_s) = \int_{\Omega_o} (\delta q_x)^2 + (c - q_s)(\delta q)^2 \; dx, \tag{6.89}$$

for the periodic solutions, i.e., the integrands in the second variations for both the soliton and periodic solutions are identical in form. We may interpret the integration variable in (6.89) as the co-moving variable provided we understand the translation is modulo $2x_o$ which is the length of the periodic domain Ω_0.

In some respects the case here in which $(c - q_s)$ is not definite is similar to the situation described in Chapter 5 for the rightward-travelling modon in which the coefficient Φ_s' takes on different signs in the interior and exterior regions. The fact that coefficients such as these in the second variation of the functional associated with the appropriate variational principle are not definite *does not* imply instability. Rather, as this chapter will make clear, the demonstration of stability (assuming it holds) will require a somewhat deeper insight into the mathematical structure of the quadratic form contained in the integrand associated with $\delta^2\mathcal{H}(q_s)$.

A critical aspect of a formal stability argument is that, of course, $\delta^2\mathcal{H}(q_s)$ is an invariant of the linear stability equation (6.86).

Theorem 6.4 $\delta^2 \mathcal{H}(q_s)$ *given by (6.87) satisfies*

$$\frac{d\delta^2 \mathcal{H}(q_s)}{dt} = 0,$$

for $\delta q(x,t)$ satisfying (6.86).

Proof. We proceed directly. From (6.89) we have

$$\frac{d\delta^2 \mathcal{H}(q_s)}{dt} = \int_{\Omega_\infty} 2\delta q_x \delta q_{xt} + 2(c - q_s)\,\delta q \delta q_t \; dx$$

$$= \int_{\Omega_\infty} 2\left[(c - q_s)\,\delta q - \delta q_{xx}\right]\delta q_t \; dx$$

$$= 2 \int_{\Omega_\infty} \left[(q_s - c)\,\delta q + \delta q_{xx}\right]\left[(q_s - c)\,\delta q + \delta q_{xx}\right]_x \; dx$$

$$= \int_{\Omega_\infty} \left[\left((q_s - c)\,\delta q + \delta q_{xx}\right)^2\right]_x \; dx = 0. \quad \blacksquare$$

Normal mode stability

The first mathematical evidence that the *KdV* soliton was stable was established by Jeffery & Kakutani (1970, 1972) who showed that there are no unstable normal mode solutions to the linear stability problem (6.86). Although the argument does not directly exploit the underlying Hamiltonian structure, it is informative and thus we include it here.

If we assume a solution to (6.86) in the form

$$\delta q(x,t) = \psi(\zeta) \exp\left(c^{\frac{3}{2}}\sigma t/8\right), \tag{6.90}$$

where $\zeta = \sqrt{c}\,x/2$ and σ is a scaled growth rate, it follows that ψ satisfies the ordinary differential equation

$$\psi''' + 4\left[3\operatorname{sech}^2(\zeta) - 1\right]\psi' + \left[\sigma - 24\operatorname{sech}^2(\zeta)\tanh(\zeta)\right]\psi = 0, \tag{6.91}$$

where the prime indicates differentiation with respect to the variable ζ. The appropriate boundary conditions on $\psi(\zeta)$ are that it smoothly vanishes at infinity. A solution of the form (6.90) implies instability if there exists a solution to (6.91) for which the real part of σ is greater than zero.

We must find three linearly independent solutions to (6.91) for a given value of σ. Let us begin, following Jeffery & Kakutani (1970, 1972), by introducing the dependent variable transformation

$$\psi\left(\zeta\right) = \psi_i\left(\zeta\right) = F_i\left(\zeta\right)\exp\left(\lambda_i\zeta\right), \tag{6.92}$$

where $\lambda_i \in (\lambda_1, \lambda_2, \lambda_3)$ is (it is assumed for now) a distinct root of the cubic equation

$$\lambda^3 - 4\lambda + \sigma = 0. \tag{6.93}$$

Substitution of (6.92), assuming (6.93), into (6.91) leads to

$$F_i''' + 3\lambda_i F_i'' + \left[3\lambda_i^2 + 12\text{sech}^2\left(\zeta\right) - 4\right]F_i'$$

$$+\text{sech}^2\left(\zeta\right)\left[12\lambda_i - 24\tanh\left(\zeta\right)\right]F_i = 0. \tag{6.94}$$

Equation (6.94) possesses a polynomial solution (with respect to $\tanh\left(\zeta\right)$) proportional to

$$F_i = \lambda_i\left(\lambda_i^2 - 4\right) + 4\left(2 - \lambda_i^2\right)\tanh\left(\zeta\right)$$

$$+8\lambda_i\tanh^2\left(\zeta\right) - 8\tanh^3\left(\zeta\right), \tag{6.95}$$

for $i = 1, 2$ and 3.

This procedure will not yield three linearly independent solutions in two cases. The first case arises when the growth rate satisfies $\sigma = 0$, i.e., $\lambda = -2$, 0 or $+2$. In this case it is straightforward to show that *all* the $\psi_i\left(\zeta\right)$ functions constructed using the above method are proportional to $\text{sech}^2\left(\zeta\right)\tanh\left(\zeta\right)$.

Modes for which the (real part of the) growth rate satisfies $\sigma = 0$ correspond to neutrally stable modes. Since the underlying dynamics is inviscid, neutral stability is the only kind of modal stability possible. Indeed, the objective of the present analysis is to show that it is only a neutrally stable mode which can satisfy the appropriate boundary conditions at infinity.

To obtain the three linearly independent solutions corresponding to $\sigma = 0$, it is convenient to go back to (6.91) and re-write it in the form

$$\frac{d}{d\zeta}\left\{\psi'' + \left[12\text{sech}^2\left(\zeta\right) - 4\right]\psi\right\} = 0,$$

which can be immediately integrated to yield

$$\psi'' + \left[12\text{sech}^2\left(\zeta\right) - 4\right]\psi = \beta_1, \tag{6.96}$$

where β_1 is a constant of integration.

Further progress is facilitated with introducing the change of independent variable

$$\varsigma = \tanh(\zeta),$$

which allows (6.96) to be re-cast in the form

$$\frac{d}{d\varsigma}\left[(1-\varsigma^2)\frac{d\psi}{d\varsigma}\right] + \left(12 - \frac{4}{1-\varsigma^2}\right)\psi = \frac{\beta_1}{1-\varsigma^2}. \tag{6.97}$$

Observing that a homogeneous solution to (6.97) is the associated Legendre function $P_3^2(\varsigma) = \varsigma(1-\varsigma^2)$ (note that in terms of the variable ζ this is just $\text{sech}^2(\zeta)\tanh(\zeta)$), allows one to construct a variation of parameters solution to (6.97) in the form

$$\psi = \varsigma(1-\varsigma^2)\,\eta(\varsigma). \tag{6.98}$$

Substitution of (6.98) into (6.97) leads to

$$\frac{d}{d\varsigma}\left[\varsigma^2(1-\varsigma^2)^3\frac{d\eta}{d\varsigma}\right] = \beta_1\varsigma. \tag{6.99}$$

Equation (6.99) can be easily integrated twice to give

$$\eta = \beta_3 + \frac{\beta_1}{16}\left[\frac{2\varsigma}{(1-\varsigma^2)^2} + \frac{3\varsigma}{1-\varsigma^2} + \frac{3}{2}\ln\left(\frac{1+\varsigma}{1-\varsigma}\right)\right]$$

$$+\beta_2\left[\frac{\varsigma}{4(1-\varsigma^2)^2} + \frac{7\varsigma}{8(1-\varsigma^2)} + \frac{15}{16}\ln\left(\frac{1+\varsigma}{1-\varsigma}\right) - \frac{1}{\varsigma}\right], \tag{6.100}$$

where β_2 and β_3 are additional constants of integration.

If (6.100) is substituted back into (6.98) and the resulting expression written in terms of the original ζ variable, it follows that in the case where $\sigma = 0$, three linearly independent solutions to (6.91) can be written in the form

$$\psi_1(\zeta) = \tanh^2(\zeta) - 2\text{sech}^2(\zeta) + 3\zeta\psi_3(\zeta), \tag{6.101}$$

$$\psi_2(\zeta) = 2\sinh^2(\zeta) + 7\tanh^2(\zeta) - 8\text{sech}^2(\zeta) + 15\zeta\psi_3(\zeta), \tag{6.102}$$

$$\psi_3(\zeta) = \text{sech}^2(\zeta)\tanh(\zeta). \tag{6.103}$$

The second situation where the ψ_i determined by (6.92) and (6.95) will *not* be linearly independent is when $\sigma = \pm 16/\sqrt{27}$. In these cases, (6.93) can be factored into

$$\left(\lambda - \frac{2}{\sqrt{3}}\right)^2\left(\lambda + \frac{4}{\sqrt{3}}\right) = 0,$$

$$\left(\lambda + \frac{2}{\sqrt{3}}\right)^2 \left(\lambda - \frac{4}{\sqrt{3}}\right) = 0,$$

respectively. Hence (6.93) has a double root at $\lambda = \pm 2/\sqrt{3}$ and a simple root at $\lambda = \mp 4/\sqrt{3}$ for $\sigma = \pm 16/\sqrt{27}$, respectively. Since the two $\psi_i\,(\zeta)$ functions, as determined by (6.92) and (6.95), associated with a specific double root will be obviously identical, it follows, of course, that they are linearly dependent.

We can find three linearly independent solutions for the double root cases as follows. Let us consider the case $\sigma = 16/\sqrt{27}$. The case corresponding to $\sigma = -16/\sqrt{27}$ is treated similarly. If we denote, say, $\lambda_{1,2} = 2/\sqrt{3}$ and $\lambda_3 = -4/\sqrt{3}$, we can choose $\psi_1\,(\zeta)$ and $\psi_3\,(\zeta)$ to be given by (6.92) and (6.95) for λ given by $2/\sqrt{3}$ and $-4/\sqrt{3}$, respectively.

To obtain a linearly independent $\psi_2\,(\zeta)$ function in the form (6.92), in the case where $\lambda_{1,2} = 2/\sqrt{3}$, we rewrite (6.94) as

$$F_2''' + 2\sqrt{3}F_2'' + \left[12\mathrm{sech}^2\,(\zeta)\,F_2\right]' + 8\sqrt{3}\mathrm{sech}^2\,(\zeta)\,F_2 = 0, \qquad (6.104)$$

and introduce the variable

$$\varsigma = \tanh\,(\zeta), \qquad (6.105)$$

which allows (6.104) to be re-written as

$$\left(1 - \varsigma^2\right)^2 \frac{d^3 F_2}{d\varsigma^3} + 2\sqrt{3}\left(1 - \varsigma^2\right)\left(1 - \sqrt{3}\varsigma\right)\frac{d^2 F_2}{d\varsigma^2}$$

$$+ \left[6\left(1 - \varsigma^2\right) + 4\left(1 - \sqrt{3}\varsigma\right)\right]\frac{dF_2}{d\varsigma} + 8\sqrt{3}\left(1 - \sqrt{3}\varsigma\right)F_2 = 0. \qquad (6.106)$$

Equation (6.106) may be solved using the method of variation of parameters in the form

$$F_2\,(\varsigma) = \Phi\,(\varsigma)\,F_1\,(\varsigma), \qquad (6.107)$$

where we write

$$F_1\,(\varsigma) = \left(2 - \sqrt{3}\varsigma\right)\left(1 - 3\varsigma^2\right),$$

which follows from (6.95) for $\lambda_1 = 2/\sqrt{3}$. After some algebra, one finds

$$\Phi\,(\varsigma) = \frac{3\varsigma}{\left(1 - 3\varsigma^2\right)} + \frac{1}{2}\ln\left(\frac{1 + \varsigma}{1 - \varsigma}\right). \qquad (6.108)$$

Thus, in summary, in the case where $\sigma = 16/\sqrt{27}$, the three linearly independent solutions to (6.91) can be written in the form, respectively,

$$\psi_1(\zeta) = \exp\left(2\zeta/\sqrt{3}\right)\left[2 - \sqrt{3}\tanh(\zeta)\right]\left[1 - 3\tanh^2(\zeta)\right], \qquad (6.109)$$

$$\psi_2(\zeta) = \exp\left(2\zeta/\sqrt{3}\right)\left[2 - \sqrt{3}\tanh(\zeta)\right] \times$$

$$\left\{3\tanh(\zeta) + \zeta\left[1 - 3\tanh^2(\zeta)\right]\right\}, \qquad (6.110)$$

$$\psi_3(\zeta) = \exp\left(-4\zeta/\sqrt{3}\right) \times$$

$$\left[5 + 5\sqrt{3}\tanh(\zeta) - 12\tanh^2(\zeta) + 3\sqrt{3}\tanh^3(\zeta)\right]. \qquad (6.111)$$

As mentioned above, the case where $\lambda_{1,2} = -2/\sqrt{3}$ and $\lambda_3 = 4/\sqrt{3}$ can be treated similarly. As before, we can choose $\psi_1(\zeta)$ and $\psi_3(\zeta)$ to be given by (6.92) and (6.95) for λ given by $-2/\sqrt{3}$ and $4/\sqrt{3}$, respectively. In this case, (6.94) can be re-written as

$$F_2''' - 2\sqrt{3}F_2'' + \left[12\,\text{sech}^2(\zeta)\,F_2\right]' - 8\sqrt{3}\,\text{sech}^2(\zeta)\,F_2 = 0, \qquad (6.112)$$

and introducing the variable (6.105) leads to

$$\left(1 - \varsigma^2\right)^2\frac{d^3F_2}{d\varsigma^3} - 2\sqrt{3}\left(1 - \varsigma^2\right)\left(1 + \sqrt{3}\varsigma\right)\frac{d^2F_2}{d\varsigma^2}$$

$$+ \left[6\left(1 - \varsigma^2\right) + 4\left(1 + \sqrt{3}\varsigma\right)\right]\frac{dF_2}{d\varsigma} - 8\sqrt{3}\left(1 + \sqrt{3}\varsigma\right)F_2 = 0, \qquad (6.113)$$

which may be solved using the method of variation of parameters in the form

$$F_2(\varsigma) = \Phi(\varsigma)\,F_1(\varsigma), \qquad (6.114)$$

where we write

$$F_1(\varsigma) = \left(2 + \sqrt{3}\varsigma\right)\left(1 - 3\varsigma^2\right),$$

which follows from (6.95) for $\lambda_1 = -2/\sqrt{3}$. It turns out, however, that $\Phi(\varsigma)$ is given by (6.108) in this case as well.

Thus, in summary, in the case where $\sigma = -16/\sqrt{27}$, the three linearly independent solutions to (6.91) can be written in the form, respectively,

$$\psi_1(\zeta) = \exp\left(-2\zeta/\sqrt{3}\right)\left[2 + \sqrt{3}\tanh(\zeta)\right]\left[1 - 3\tanh^2(\zeta)\right], \qquad (6.115)$$

$$\psi_2(\zeta) = \exp\left(-2\zeta/\sqrt{3}\right)\left[2 + \sqrt{3}\tanh(\zeta)\right] \times$$

$$\left\{3\tanh(\zeta) + \zeta\left[1 - 3\tanh^2(\zeta)\right]\right\}, \qquad (6.116)$$

$$\psi_3(\zeta) = \exp\left(4\zeta/\sqrt{3}\right) \times$$

$$\left[-5 + 5\sqrt{3}\tanh(\zeta) + 12\tanh^2(\zeta) + 3\sqrt{3}\tanh^3(\zeta)\right]. \qquad (6.117)$$

We now argue that the only solution to the linear stability problem which satisfies the boundary conditions at infinity is a $\sigma = 0$ solution, i.e., the *KdV* soliton is neutrally stable to normal mode perturbations. Consider first the case where we have three distinct roots to (6.93). If any of the roots possess the property that the real part is greater than or equal to zero, i.e., $\text{Re}(\lambda_i) \geq 0$, the corresponding $\psi_i(\zeta)$, as determined by (6.92) and (6.95), will not vanish as $\zeta \to \infty$ unless $F_i(\zeta)$ does. But this can only occur if $\lambda(\lambda - 2)^2 = 0$, i.e., $\lambda = 0$ or 2. But these values of λ imply that $\sigma = 0$. Indeed, it is straightforward to verify, as mentioned above, that $\psi_i(\zeta)$, as determined by (6.92) and (6.95), is proportional to (6.103). Likewise, if $\text{Re}(\lambda_i) < 0$, the corresponding $\psi_i(\zeta)$, as determined by (6.92) and (6.95), will not vanish as $\zeta \to -\infty$ unless $\lambda = 0$ or -2. But this also implies that $\sigma = 0$ and that $\psi_i(\zeta)$, as determined by (6.92) and (6.95), is proportional to (6.103). It is clear that of the $\sigma = 0$ solutions, as determined by (6.101), (6.102) and (6.103), only (6.103) will vanish at infinity. In addition, none of the solutions associated with $\sigma = 16/\sqrt{27}$, as determined by (6.109), (6.110) and (6.111), or with $\sigma = -16/\sqrt{27}$, as determined by (6.116), (6.117) and (6.118), will vanish at infinity. Finally, it is straightforward to show that any linear combination of the above solutions which do not vanish at infinity does not itself vanish at infinity for all time (when the time dependence is included; see (6.90)). Thus we conclude that the only acceptable solution to the normal mode stability problem (6.91) is the neutrally stable one (6.103).

Implications for the general stability problem

Since (6.103) has $\sigma = 0$ associated with it, it follows that (6.103) in fact corresponds to an exact steady solution of the original linear stability equation (6.86). Indeed, this solution has an immediate physical interpretation. If (6.103) is substituted back into (6.90), it follows that

$$\delta q\left(x,t\right) = \mathrm{sech}^2\left(\sqrt{c}\, x/2\right)\tanh\left(\sqrt{c}\, x/2\right), \tag{6.118}$$

where, for convenience, we have not included the free multiplicative constant which is allowed. But this is proportional simply to the principal increment associated with a translation of the *KdV* soliton in the x direction. From

$$q_s\left(x\right) = 3c\,\mathrm{sech}^2\left[\frac{\sqrt{c}\, x}{2}\right], \tag{6.119}$$

it follows that

$$\Delta q_s\left(x\right) \equiv q_s\left(x + \Delta x\right) - q_s\left(x\right) \simeq q_{s_x}\left(x\right)\Delta x + O\left(\left(\Delta x\right)^2\right)$$

$$= 3\sqrt{3}c\,\mathrm{sech}^2\left(\sqrt{c}\, x/2\right)\tanh\left(\sqrt{c}\, x/2\right)\Delta x + O\left(\left(\Delta x\right)^2\right). \tag{6.120}$$

Thus, the solution (6.103) corresponds to an infinitesimal x-translation of the soliton, given by,

$$\delta q\left(x,t\right) = q_{s_x}\left(x\right). \tag{6.121}$$

We call this solution to the linear stability problem the *translational mode*.

The translational mode corresponds to an element of the kernel of the operator $\mathcal{L}$ defined by (6.88), i.e.,

$$\mathcal{L}q_{s_x}\left(x\right) = \left[-\partial_{xx} + c - q_s\left(x\right)\right]q_{s_x}\left(x\right) = 0, \tag{6.122}$$

which, in turn, implies

$$\delta^2\mathcal{H}\left(q_s\right)\big|_{\delta q = q_{s_x}} = 0. \tag{6.123}$$

While these facts can be verified by direct calculation, (6.122) is just (6.77) written with respect to $\tilde{x}$ (with the tilde deleted) and (6.123) follows from multiplying (6.77) by q_{s_x} and integrating the result over Ω_∞ giving

$$0 = \int_{\Omega_\infty} q_{s_x}q_{s_{xxx}} + \left(q_s - c\right)\left(q_{s_x}\right)^2 \, dx = -\int_{\Omega_\infty} q_{s_x}\mathcal{L}q_{s_x}\, dx$$

$$= -\int_{\Omega_\infty}\left(q_{s_{xx}}\right)^2 + \left(c - q_s\right)\left(q_{s_x}\right)^2 \, dx$$

$$= -\,\delta^2\mathcal{H}\left(q_s\right)\big|_{\delta q = q_{s_x}} \tag{6.124}$$

It is possible to interpret (6.124) as "Andrews's Theorem" in the context of the *KdV* equation.

This last result has important consequences in proving formal or linear stability in the sense of Liapunov. In order to bound $\delta^2 \mathcal{H}(q_s)$ away from zero in terms of a suitable perturbation norm we will have to factor out, in an appropriate mathematical manner, all perturbations which correspond to a simple translation of the soliton. As shown by Benjamin (1972) (see also Benjamin *et al.*, 1972), and we describe below, this can be accomplished by the introduction of a "sliding" pseudo-metric on the perturbation field.

We will have to factor arbitrary translations in the development of the nonlinear stability theory as well. A simple example will serve to illustrate the problem. It is straightforward to verify that the solution to (6.84) subject to the initial condition

$$q(x,0) = 3(c+\delta)\operatorname{sech}^2\left(\frac{\sqrt{c+\delta}}{2}x\right), \qquad (6.125)$$

is given by

$$q(x,t) = 3(c+\delta)\operatorname{sech}^2\left(\frac{\sqrt{c+\delta}}{2}(x-\delta t)\right), \qquad (6.126)$$

where δ is any real number satisfying $c + \delta > 0$. Comparing (6.127) to (6.119) we see that this solution is also simply a soliton but with the translation velocity given by $c + \delta$ compared to c in $q_s(x)$. We remind the reader that both solutions are written with respect to the co-moving frame of reference associated with (6.84) which accounts for the δt term in the argument of sech (∗) in (6.126).

Although (6.126) is valid for all $\delta > -c$, from the point of view of the stability theory we may consider $|\delta|$ as a small but finite positive real number. The initial condition (6.125) can therefore be thought of as a small but finite amplitude distortion of the soliton (6.119).

The $L_2(\Omega_\infty)$ norm of the initial disturbance, or perturbation, field, given by

$$\|q(x,0) - q_s(x)\| \equiv \left\{\int_{-\infty}^{\infty} [q(x,0) - q_s(x)]^2 \, dx\right\}^{\frac{1}{2}}, \qquad (6.127)$$

can be made as small as one wants by controlling the size of $|\delta|$ since

$$\|q(x,0) - q_s(x)\|^2 = 9\int_{-\infty}^{\infty}\left\{(c+\delta)^2\operatorname{sech}^4\left(\frac{\sqrt{c+\delta}}{2}x\right)\right.$$

$$\left. + c^2\operatorname{sech}^4\left(\frac{\sqrt{c}}{2}x\right)\right\} \, dx$$

$$-18c\left(c+\delta\right)\int_{-\infty}^{\infty}\operatorname{sech}^{2}\left(\frac{\sqrt{c+\delta}}{2}x\right)\operatorname{sech}^{2}\left(\frac{\sqrt{c}}{2}x\right)\,dx$$

$$\leq 48\sqrt[3]{\gamma_{1}}-18\gamma_{2}^{2}\int_{-\infty}^{\infty}\operatorname{sech}^{4}\left(\sqrt{\gamma_{2}}\,x/2\right)\,dx$$

$$=48\left(\sqrt[3]{\gamma_{1}}-\sqrt[3]{\gamma_{2}}\right)\leq\frac{48\left|\delta\right|\left(3c^{2}+3c\delta+\delta^{2}\right)}{\sqrt[3]{c}},\tag{6.128}$$

where $\gamma_{1}=\max\left(c+\delta,c\right)>0$ and $\gamma_{2}=\min\left(c+\delta,c\right)>0$. Thus $\|q\left(x,0\right)-q_{s}\left(x\right)\|\to 0$ as $|\delta|\to 0$.

However, and this is the point, as time increases the bulk of the support of $q\left(x,t\right)$ translates away from the origin while the bulk of the support of $q_{s}\left(x\right)$ remains localized around the origin. Thus, after a sufficiently large time, the overlap between $q\left(x,t\right)$ and $q_{s}\left(x\right)$ consists of exponentially small and monotonically decreasing values so that

$$\|q\left(x,t\right)-q_{s}\left(x\right)\|^{2}\to\|q\left(x,t\right)\|^{2}+\|q_{s}\left(x\right)\|^{2},$$

as $t\to\infty$ which cannot be made arbitrarily small and is, indeed, bounded away from 0 no matter the size of $|\delta|$, i.e., $\|q\left(x,0\right)-q_{s}\left(x\right)\|$.

This can be seen from

$$\|q\left(x,t\right)-q_{s}\left(x\right)\|^{2}=9\int_{-\infty}^{\infty}\left\{\left(c+\delta\right)^{2}\operatorname{sech}^{4}\left(\frac{\sqrt{c+\delta}}{2}\left(x-\delta t\right)\right)\right.$$

$$\left.+c^{2}\operatorname{sech}^{4}\left(\frac{\sqrt{c}}{2}\,x\right)\right\}\,dx$$

$$-18c\left(c+\delta\right)\int_{-\infty}^{\infty}\operatorname{sech}^{2}\left(\frac{\sqrt{c+\delta}}{2}\left(x-\delta t\right)\right)\operatorname{sech}^{2}\left(\frac{\sqrt{c}}{2}x\right)\,dx$$

$$=24\left[\left(c+\delta\right)^{\frac{3}{2}}+c^{\frac{3}{2}}\right]-18c\left(c+\delta\right)\times$$

$$\int_{-\infty}^{\infty}\operatorname{sech}^{2}\left(\frac{\sqrt{c+\delta}}{2}\left(x-\delta t\right)\right)\operatorname{sech}^{2}\left(\frac{\sqrt{c}}{2}x\right)\,dx.\tag{6.129}$$

However since,

$$\operatorname{sech}^{2}\left(\frac{\sqrt{c+\delta}}{2}\left(x-\delta t\right)\right)=\frac{\operatorname{sech}^{2}\left(\frac{\sqrt{c+\delta}}{2}x\right)\operatorname{sech}^{2}\left(\frac{\sqrt{c+\delta}}{2}\delta t\right)}{\left[1-\tanh\left(\frac{\sqrt{c+\delta}}{2}x\right)\tanh\left(\frac{\sqrt{c+\delta}}{2}\delta t\right)\right]^{2}}$$

$$= \frac{\text{sech}^2\left(\frac{\sqrt{c+\delta}}{2}x\right)\text{sech}^2\left(\frac{\sqrt{c+\delta}}{2}\delta t\right)\left[1+\tanh\left(\frac{\sqrt{c+\delta}}{2}x\right)\tanh\left(\frac{\sqrt{c+\delta}}{2}\delta t\right)\right]^2}{1-\tanh^2\left(\frac{\sqrt{c+\delta}}{2}x\right)\tanh^2\left(\frac{\sqrt{c+\delta}}{2}\delta t\right)}$$

$$\leq \text{sech}^2\left(\frac{\sqrt{c+\delta}}{2}\delta t\right)\left[1+\tanh\left(\frac{\sqrt{c+\delta}}{2}x\right)\tanh\left(\frac{\sqrt{c+\delta}}{2}\delta t\right)\right]^2$$

$$\leq 4\text{sech}^2\left(\frac{\sqrt{c+\delta}}{2}\delta t\right),$$

so that

$$24\left[(c+\delta)^{\frac{3}{2}}+c^{\frac{3}{2}}\right] \geq \|q(x,t)-q_s(x)\|^2 \geq 24\left[(c+\delta)^{\frac{3}{2}}+c^{\frac{3}{2}}\right]$$

$$-72c(c+\delta)\,\text{sech}^2\left(\frac{\sqrt{c+\delta}}{2}\delta t\right)\int_{-\infty}^{\infty}\text{sech}^2\left(\sqrt{c}\,x/2\right)\,dx$$

$$= 24\left[\sqrt[3]{c+\delta}+\sqrt[3]{c}\right] - 288\sqrt{c}\,(c+\delta)\,\text{sech}^2\left(\frac{\sqrt{c+\delta}}{2}\delta t\right), \qquad (6.130)$$

which implies

$$\lim_{t\to\infty}\|q(x,t)-q_s(x)\|^2 = 24\left[\sqrt[3]{c+\delta}+\sqrt[3]{c}\right]$$

$$= \|q(x,t)\|^2 + \|q_s(x)\|^2, \qquad (6.131)$$

which cannot be made arbitrarily small by controlling $|\delta|$.

This example illustrates that we must be careful how we define stability for the soliton. Here we imposed a small initial perturbation to the soliton $q_s(x)$ by changing its amplitude and hence its speed. This results in a translation of the perturbed soliton relative to the unperturbed soliton which, over time, implies that the $L_2(\Omega_\infty)$ norm of the difference between the perturbed and unperturbed soliton cannot be made arbitrarily small, i.e., for every $\varepsilon > 0$ it is *not* possible to find a $\delta > 0$ such that

$$\|q(x,0)-q_s(x)\| < \delta \Rightarrow \|q(x,t)-q_s(x)\| < \varepsilon,$$

for all $t \geq 0$ with respect to the $L_2(\Omega_\infty)$ norm (6.127). The resolution to this issue is described in the next section.

Finally, we remark that the nonlinearity in the *KdV* equation does not make significantly more complex the underlying stability analysis once we have overcome the issues just raised. Thus there is little point in splitting our work into one section dealing explicitly with the formal and linear stability theory and another with the nonlinear stability theory. Our approach will be to present the nonlinear theory directly.

6.6   Nonlinear stability

As shown by Benjamin (1972) (see also Benjamin *et al.*, 1972 and Bona, 1975), the stability of the KdV soliton can be established using the "sliding" pseudo-metric defined by

$$d\left(f,g\right) \equiv \inf_{y \in \mathbb{R}} \left\| f\left(x\right) - g\left(x+y\right) \right\|_1, \tag{6.132}$$

where the norm $\left\| f \right\|_1$ is given by

$$\left\| f \right\|_1 \equiv \left\{ \int_{-\infty}^{\infty} \left[f_x\left(x\right) \right]^2 + \left[f\left(x\right) \right]^2 \, dx \right\}^{\frac{1}{2}}, \tag{6.133}$$

where $f\left(x\right)$ and $g\left(x\right)$ are arbitrary continuously differentiable functions which possess the property that they and their first derivatives are in $L_2\left(\Omega_\infty\right)$.

The metric $d\left(f,g\right)$ is called a pseudo-metric because $d\left(f,g\right) = 0$ does not necessarily imply $f = g$. One can immediately see the advantage of introducing this metric in that it allows one to factor out translations in the perturbed soliton. That is, f will be close to g if there exists a translation for which the translated g is close to f in the $\left\| * \right\|_1$ norm. In the stability theory to come, we will work with both $\left\| * \right\|_1$ and the $L_2\left(\Omega_\infty\right)$ norm $\left\| * \right\|$. Obviously, we have

$$\left\| f \right\| \leq \left\| f \right\|_1. \tag{6.134}$$

The norm $\left\| f \right\|_1$ is the standard norm associated with the Sobolev space of $L_2\left(\Omega_\infty\right)$ functions whose (generalized) first derivatives are also in $L_2\left(\Omega_\infty\right)$. We denote this function space $H^1\left(\Omega_\infty\right)$ and note that it is a Hilbert space for the scalar product

$$\left(f,g\right)_1 = \int_{-\infty}^{\infty} f_x g_x + fg \, dx,$$

so that

$$\left\| f \right\|_1 = \sqrt{\left(f,f\right)_1}.$$

We point out that sometimes (e.g., Benjamin, 1972 or Ladyzhenskaya, 1969) the function space $H^1\left(\Omega_\infty\right)$ is denoted as $W^{1,2}\left(\Omega_\infty\right)$.

The above generalizes in a straightforward way so that $H^k\left(\Omega_\infty\right)$ (or $W^{k,2}\left(\Omega_\infty\right)$) is the Sobolev space of $L_2\left(\Omega_\infty\right)$ functions whose (generalized) derivatives up to order k are also in $L_2\left(\Omega_\infty\right)$. The standard inner product associated with $H^k\left(\Omega_\infty\right)$ is given by

$$\left(f,g\right)_k = \int_{-\infty}^{\infty} \sum_{m=0}^{k} \left(\frac{d^m f}{dx^m}\right) \left(\frac{d^m g}{dx^m}\right) \, dx,$$

with the norm

$$\|f\|_k = \sqrt{(f,f)_k}.$$

In this notation

$$L_2\left(\Omega_\infty\right) = H^0\left(\Omega_\infty\right) = W^{0,2}\left(\Omega_\infty\right),$$

and for convenience

$$(f,g) \equiv (f,g)_0 \text{ and } \|f\| \equiv \|f\|_0.$$

The intersection $\cap_{k=0}^\infty H^k\left(\Omega_\infty\right)$ is denoted as $H^\infty\left(\Omega_\infty\right)$ which is the Sobolev space of smooth $L_2\left(\Omega_\infty\right)$ functions with the property that all of its derivatives are also in $L_2\left(\Omega_\infty\right)$. The reader is referred to, for example, Adams (1975) for the theory of Sobolev spaces and to, for example, Ladyzhenskaya (1969) and Temam (1983) for a description of their application to fluid mechanics. The functional analysis of Sobolev spaces which we require is not overly sophisticated and will be developed as we go along.

We are now in a position where it is possible for us to give a clear statement, as a theorem, of what "stability" means for the KdV soliton.

Theorem 6.5 *The KdV soliton, given by (6.119), is stable in the sense that for every $\varepsilon > 0$ there exists $\delta > 0$ such that*

$$\|q(x,0) - q_s\left(x\right)\|_1 < \delta \Rightarrow d\left(q\left(x,t\right), q_s\left(x\right)\right) < \varepsilon,$$

for all $t > 0$ where $q\left(x,t\right)$ solves (6.84) with initial data $q(x,0)$ where $q\left(x,t\right) \in H^\infty\left(\Omega_\infty\right)$ for all $t \geq 0$.

Our objective for the remainder of this section is to prove this theorem. The approach we take will follow closely the argument presented by Albert *et al.* (1987). Although we will focus on the stability argument as it pertains to the KdV equation (6.84), the stability argument for solitary wave solutions of various nonlinear wave equations is similar in spirit, if not in detail, to that presented here. For example, the stability argument presented by Albert *et al.* (1987) is directly applicable to certain generalized KdV, intermediate or regularized long-wave and Benjamin-Ono equations (see also, for example, Bennett *et al.*, 1983 and Weinstein, 1986). (We do not describe these equations here. The interested reader should consult Drazin & Johnson (1987), Newell (1985) or Ablowitz & Segur (1981) for a description of these models.) We also explicitly mention the work of Bona *et al.* (1987) which establish *necessary and sufficient* conditions for the stability (and hence instability) of solitary wave solutions for generalized KdV and Benjamin-Ono equations. A relatively recent survey of solitary wave stability theory is given by Bona & Soyeur (1994).

We begin our work by introducing the functional

$$\Delta \mathcal{H} = 2\left[\mathcal{H}\left(q\left(x,t\right)\right) - \mathcal{H}\left(q_s\right)\right], \qquad (6.135)$$

where $\mathcal{H}\left(*\right)$ is given by (6.80) and where $q\left(x,t\right)$ solves the KdV equation (6.84). By construction

$$\frac{d\left[\Delta \mathcal{H}\right]}{dt} = 0, \qquad (6.136)$$

since both $\mathcal{H}\left(q\left(x,t\right)\right)$ and $\mathcal{H}\left(q_s\right)$ are themselves individually invariant with respect to time. The initial condition for $q\left(x,t\right)$ is given by $q\left(x,0\right) = q_0\left(x\right)$. We assume that $q\left(x,t\right) \in H^1\left(\Omega_\infty\right)$ for $t \geq 0$ (the reader is referred to Benjamin *et al.*, 1972 and Bona & Smith, 1975 for the relevant existence and regularity theory for the initial value problem).

Let us introduce the function $\varphi\left(x,t\right)$ defined by

$$\varphi\left(x,t\right) = q\left(x,t\right) - q_s\left(x + a\left(t\right)\right). \qquad (6.137)$$

The function $a\left(t\right)$ will be chosen in our subsequent work and is an important aspect in the development of the stability theory. Note that it follows from (6.132) that

$$d\left(q\left(x,t\right), q_s\left(x\right)\right) \leq \|\varphi\|_1, \qquad (6.138)$$

regardless of the choice of $a\left(t\right)$ so that, roughly speaking, if we can choose $a\left(t\right)$ to make $\|\varphi\|_1$ sufficiently small for $t \geq 0$, stability with respect to the sliding metric (6.132) will be ensured.

It is convenient to write $\Delta \mathcal{H}$ in terms of φ. If (6.80) is used explicitly, it follows that $\Delta \mathcal{H}$ may be written in the form

$$\Delta \mathcal{H}\left(\varphi\right) = \int_{\Omega_\infty} \left(\varphi_x\right)^2 + \left[c - q_s\left(x + a\left(t\right)\right)\right]\varphi^2 - \frac{\varphi^3}{3} \, dx, \qquad (6.139)$$

or, equivalently, as

$$\Delta \mathcal{H}\left(\varphi\right) = \int_{\Omega_\infty} \varphi \mathcal{L}_a \varphi - \frac{\varphi^3}{3} \, dx, \qquad (6.140)$$

where $\mathcal{L}_a$ is given by

$$\mathcal{L}_a = -\partial_{xx} + c - q_s\left(x + a\left(t\right)\right). \qquad (6.141)$$

In order to bound $\Delta \mathcal{H}\left(\varphi\right)$ in terms of $\|\varphi\|_1$, we need the estimate

Lemma 6.6 *If $\varphi\left(x,t\right) \in H^1\left(\Omega_\infty\right)$, then*

$$\int_{\Omega_\infty} \varphi^3 dx \leq \frac{1}{\sqrt{2}} \|\varphi\|_1^3. \qquad (6.142)$$

Proof. If $\varphi(x,t) \in H^1(\Omega_\infty)$, then the Fourier Transform of $\varphi(x,t)$, denoted as $\hat{\varphi}(k,t)$, exists and is related to $\varphi(x,t)$ via the Fourier inversion formula

$$\varphi(x,t) = \frac{1}{\sqrt{2\pi}} \int_{-\infty}^{\infty} \hat{\varphi}(k,t) \exp(ikx) \ dk.$$

If we take the complex modulus of both sides of this expression, we have

$$|\varphi(x,t)| \le \frac{1}{\sqrt{2\pi}} \int_{-\infty}^{\infty} |\hat{\varphi}(k,t)| \ dk$$

$$= \frac{1}{\sqrt{2\pi}} \int_{-\infty}^{\infty} \frac{\sqrt{1+k^2}}{\sqrt{1+k^2}} |\hat{\varphi}(k,t)| \ dk$$

$$\le \frac{1}{\sqrt{2\pi}} \left\{ \int_{-\infty}^{\infty} (1+k^2) |\hat{\varphi}(k,t)|^2 \ dk \right\}^{\frac{1}{2}} \left\{ \int_{-\infty}^{\infty} \frac{dk}{1+k^2} \right\}^{\frac{1}{2}}$$

$$= \frac{1}{\sqrt{2}} \left\{ \int_{-\infty}^{\infty} \int_{\Omega_\infty} (\varphi_x)^2 + \varphi^2 \ dx \right\}^{\frac{1}{2}} = \frac{\|\varphi\|_1}{\sqrt{2}}, \tag{6.143}$$

where we have used the triangle and Schwarz inequalities and Parseval's Identity (see, e.g., Royden, 1968 and Zauderer, 1983). We note that since the right hand side of (6.143) is independent of x, it follows that

$$\sup_{\Omega_\infty} |\varphi(x,t)| \le \frac{\|\varphi\|_1}{\sqrt{2}}. \tag{6.144}$$

It therefore follows that

$$\int_{\Omega_\infty} \varphi^3 dx \le \int_{\Omega_\infty} |\varphi| \, \varphi^2 dx$$

$$\le \left\{ \sup_{\Omega_\infty} |\varphi(x,t)| \right\} \int_{\Omega_\infty} \varphi^2 \ dx \le \frac{\|\varphi\|_1^3}{\sqrt{2}}.$$

This completes the proof. ∎

Using (6.142) we can immediately establish the following initial (there is much more work to do) lower and upper bounds for $\Delta\mathcal{H}(\varphi)$. From (6.139) we have

$$(\varphi, \mathcal{L}_a\varphi) - \frac{\|\varphi\|_1^3}{3\sqrt{2}} \le \Delta\mathcal{H}(\varphi) \le \int_{\Omega_\infty} (\varphi_x)^2 + c\varphi^2 - \frac{\varphi^3}{3} \ dx$$

$$\leq \max (1, c) \, \|\varphi\|_1^2 + \frac{\|\varphi\|_1^3}{3\sqrt{2}}. \tag{6.145}$$

One of the major mathematical issues we have to confront is the derivation of an appropriate lower bound for $(\varphi, \mathcal{L}_a \varphi)$ in terms of $\|\varphi\|_1$. This can be done after we examine the spectrum, i.e., the eigenvalues, and the eigenfunctions of the operator $\mathcal{L}$ which we do now.

Lemma 6.7 *Consider the eigenvalue problem*

$$\mathcal{L}\psi = \lambda\psi, \tag{6.146}$$

where $x \in \Omega_\infty$ *and* $\psi \in H^1(\Omega_\infty)$ *with* $\mathcal{L}$ *given by (6.88), then the spectrum of* $\mathcal{L}$, *denoted as* spec$(\mathcal{L})$, *satisfies the following three properties:*

(P_1) *The eigenvalue* $\lambda = 0$ *is simple.*

(P_2) *The intersection of* spec$(\mathcal{L})$ *with the negative real axis consists of a single, simple eigenvalue which we denote as* λ_1.

(P_3) *If the normalized eigenfunction associated with the eigenvalue* λ_1 *is denoted as* ψ_1, *then the inequality*

$$\left(1 + \frac{\lambda_2}{|\lambda_1|}\right) \left[\frac{(q_s, \psi_1)}{\|q_s\|}\right]^2 > 1, \tag{6.147}$$

holds, where $\lambda_2 = \inf \{\lambda \in \text{spec}(\mathcal{L}) \mid \lambda > 0\}$.

In many respects, this Lemma, which is taken from Albert *et al.* (1987), is one of the most important results in this section. As shown by Albert *et al.* (1987), there are many KdV-like and Benjamin-Ono-like model equations for solitary waves for which the stability theory revolves around a functional similar to $\Delta \mathcal{H}(\varphi)$ as given by (6.140), although with a different operator $\mathcal{L}_a$ and a different nonlinearity rather than the cubic term in (6.140). Nevertheless, Albert *et al.* (1987) showed that the three properties listed in Lemma 6.7 are *sufficient* to establish the nonlinear stability of the solitary wave solutions to the above generalized class of nonlinear wave models.

Proof. The easiest way to prove Lemma 6.7 is directly. The eigenvalue problem (6.146) can be written in the form

$$\left(\partial_{xx} - c + 3c\,\text{sech}^2\left(\sqrt{c}\,x/2\right)\right)\psi = -\lambda\psi,$$

which, after introducing the change of independent variable $\xi = \tanh\left(\sqrt{c}\,x/2\right)$, leads to

$$\left(1 - \xi^2\right)\frac{d^2\psi}{d\xi^2} - 2\xi\frac{d\psi}{d\xi} + \left(12 - \frac{4 - \tilde{\lambda}}{1 - \xi^2}\right)\psi = 0, \tag{6.148}$$

where $\widetilde{\lambda} = 4\lambda/c$.

Equation (6.148) is an associated Legendre equation of degree 3 (see, e.g., Morse & Feshbach, 1963 or Arfken & Weber, 1995). The discrete spectrum consists of $\widetilde{\lambda} = -5$, 0 and 3 which implies, since $c > 0$, that $\lambda_1 = -5c/4$. The continuous spectrum consists of $\widetilde{\lambda} \geq 4$ (i.e., $\lambda \geq c$) and these solutions can be written in terms of hypergeometric functions which do not directly concern us here. It follows that and $\lambda_2 = 3c/4$.

In terms of ξ, the eigenfunctions associated with the eigenvalues λ_1, 0 and λ_2, respectively, are proportional to the associated Legendre functions

$$P_3^3\left(\xi\right) = \left(1 - \xi^2\right)^{\frac{3}{2}},$$

$$P_3^2\left(\xi\right) = \xi\left(1 - \xi^2\right),$$

$$P_3^1\left(\xi\right) = \left(5\xi^2 - 1\right)\left(1 - \xi^2\right)^{\frac{1}{2}},$$

respectively. The eigenvalues λ_1, 0 and λ_2 are obviously simple. We have therefore established properties (P_1) and (P_2). Note that written with respect to x, the eigenfunction associated with the 0 eigenvalue is proportional to $\mathrm{sech}^2\left(\sqrt{c}\,x/2\right)\tanh\left(\sqrt{c}\,x/2\right)$, i.e., the translational mode.

To establish property (P_3), we need the normalized eigenfunction $\psi_1\left(x\right)$ associated with the eigenvalue λ_1 given by

$$\psi_1\left(x\right) = \sqrt{\frac{15\sqrt{c}}{32}}\,\mathrm{sech}^3\left(\sqrt{c}\,x/2\right). \tag{6.149}$$

Thus we have

$$\left(1 + \frac{\lambda_2}{|\lambda_1|}\right)\left[\frac{(q_s, \psi_1)}{\|q_s\|}\right]^2 = \frac{3\left[\int_{-\infty}^{\infty}\mathrm{sech}^5\left(x\right)dx\right]^2}{2\int_{-\infty}^{\infty}\mathrm{sech}^4\left(x\right)dx}$$

$$= \frac{81\pi^2}{128} \simeq 6.2456 > 1.$$

This completes the proof. ■

As it turns out, it is convenient to express property (P_3) in a slightly different form. It follows from (6.147) that

$$\lambda_2 - |\lambda_1|\left[\frac{\|q_s\|^2}{(q_s, \psi_1)^2} - 1\right] > 0.$$

Because of the strict inequality, there exists a real number $\alpha > 0$ such that

$$\beta \equiv \lambda_2 - (|\lambda_1| + \alpha) \left[\frac{\|q_s\|^2}{(q_s, \psi_1)^2} - 1 \right] > 0. \tag{6.150}$$

We are now in position to begin deriving an effective lower bound for $(\varphi, \mathcal{L}_a \varphi)$. We start by establishing the following result

Lemma 6.8 *Suppose* $f(x) \in H^1(\Omega_\infty)$ *and* $a \in \mathbb{R}$ *are such that*

$$\|q_s(x + a) + f(x)\| = \|q_s\|, \tag{6.151}$$

$$\int_{\Omega_\infty} q_{s_x}(x + a) f(x) \, dx = 0, \tag{6.152}$$

then there exist positive constants A_1 *and* A_2, *independent of* $f(x)$ *and* a, *i.e., depending only on* $q_s(x)$, *such that*

$$(f, \mathcal{L}_a f) \geq A_1 \|f\|_1^2 - A_2 \left(\|f\|_1^3 + \|f\|_1^4 \right). \tag{6.153}$$

where $\mathcal{L}_a$ *is given by (6.141).*

Proof. By introducing the translation $x = \tilde{x} - a$ and $f(\tilde{x} - a) = \tilde{f}(x)$, we may, after dropping the tildes, simply set $a = 0$ and note that $\mathcal{L}_0 = \mathcal{L}$. Thus we have

$$(f, \mathcal{L}f) = \int_{-\infty}^{\infty} (f_x)^2 + [c - q_s(x)] f^2 \, dx$$

$$= \|f\|_1^2 + \int_{-\infty}^{\infty} [c - q_s(x) - 1] f^2 \, dx$$

$$\geq \|f\|_1^2 - (1 + 2c) \|f\|^2, \tag{6.154}$$

where we have used $q_s(x) \leq 3c$. This is our first estimate for a lower bound for $(f, \mathcal{L}f)$ but we will need to improve on this.

We now write $f(x)$ and $q_s(x)$ as

$$f(x) = p\psi_1(x) + f_1(x), \tag{6.155}$$

$$q_s(x) = q\psi_1(x) + q_1(x), \tag{6.156}$$

where

$$p \equiv (f, \psi_1), \quad q \equiv (q_s, \psi_1), \tag{6.157}$$

$$(f_1, \psi_1) = (q_1, \psi_1) = 0, \tag{6.158}$$

where $\psi_1(x)$ is given by (6.149) and we recall $\|\psi_1\| = 1$. In terms of the translated variables, the assumption (6.151) implies

$$2(f, q_s) + \|f\|^2 = 0,$$

which, if (6.155) and (6.156) are substituted in, yields

$$2pq + (f_1, q_1) + \|f\|^2 = 0,$$

or, equivalently,

$$p = -\frac{(f_1, q_1)}{q} - \frac{\|f\|^2}{2q}. \tag{6.159}$$

We note that property (P_3) implies $q = (q_s, \psi_1) \neq 0$.

Squaring both sides of (6.159) leads to

$$p^2 = \frac{(f_1, q_1)^2}{q^2} + \frac{(f_1, q_1)\|f\|^2}{q^2} + \frac{\|f\|^4}{4q^2}$$

$$\leq \frac{\|f_1\|^2 \|q_1\|^2}{q^2} + \frac{\|f_1\| \|q_1\| \|f\|^2}{q^2} + \frac{\|f\|^4}{4q^2}, \tag{6.160}$$

where we have used the Schwarz inequality. It follows from (6.155) through to (6.158) that

$$\left. \begin{aligned} \|f\|^2 &= p^2 + \|f_1\|^2 \geq \|f_1\|^2, \\ \frac{\|q_1\|^2}{q^2} &= \frac{\|q_s\|^2}{q^2} - 1, \end{aligned} \right\} \tag{6.161}$$

which, together with (6.160), implies

$$p^2 \leq \|f_1\|^2 \left(\frac{\|q_s\|^2}{q^2} - 1 \right) + A_3 \left(\|f\|^3 + \|f\|^4 \right), \tag{6.162}$$

where $A_3 = \max \left(\frac{\|q_1\|}{q^2}, \frac{1}{4q^2} \right) > 0$. The positive constant A_3 is obviously independent of $f(x)$ and a.

Now, substituting the decomposition (6.155) into $(f, \mathcal{L}f)$ directly, yields

$$(f, \mathcal{L}f) = (p\psi_1 + f_1, \mathcal{L}(p\psi_1 + f_1))$$

$$= -\frac{5c}{4}p^2 + p(\psi_1, \mathcal{L}f_1) + (f_1, \mathcal{L}f_1)$$

$$= -\frac{5c}{4}p^2 + (f_1, \mathcal{L}f_1),$$

since

$$(\psi_1, \mathcal{L}f_1) = (\mathcal{L}\psi_1, f_1) = -\frac{5c}{4}(\psi_1, f_1) = 0.$$

We must estimate $(f_1, \mathcal{L}f_1)$. First, we note that the assumption (6.152) implies that $f(x)$ and hence $f_1(x)$ is orthogonal to the translational mode q_{s_x} which is the eigenfunction associated with the 0 eigenvalue of the eigenvalue problem (6.146), i.e., $(q_{s_x}, f_1) = 0$, in addition to $(\psi_1, f_1) = 0$. Now, the Raleigh-Ritz variational principle for the eigenvalue $\lambda_2 = \frac{3c}{4}$ in Lemma 6.7 can be formulated (see, e.g., Zauderer, 1983) in the form

$$\lambda_2 = \min_{\psi} \left[\frac{(\psi, \mathcal{L}\psi)}{(\psi, \psi)} \right],$$

over all functions $\psi \in H^1(\Omega_\infty)$ for which $(\psi, \psi_1) = (\psi, q_{s_x}) = 0$. Clearly, $f_1(x)$ is a candidate function, so we conclude

$$(f_1, \mathcal{L}f_1) \geq \frac{3c}{4} \|f_1\|^2,$$

and hence

$$(f, \mathcal{L}f) \geq -\frac{5c}{4}p^2 + \frac{3c}{4}\|f_1\|^2. \qquad (6.163)$$

However, (6.163) can be re-written as

$$(f, \mathcal{L}f) \geq \alpha p^2 - \left(\frac{5c}{4} + \alpha\right) p^2 + \frac{3c}{4}\|f_1\|^2,$$

where α is as defined in (6.150). And using (6.162) we have

$$(f, \mathcal{L}f) \geq \alpha p^2 + \left[\frac{3c}{4} - \left(\frac{5c}{4} + \alpha\right)\left(\frac{\|q_s\|^2}{q^2} - 1\right)\right]\|f_1\|^2$$

$$-A_3 \left(\frac{5c}{4} + \alpha\right) \left(\|f\|^3 + \|f\|^4\right)$$

$$= \alpha p^2 + \beta \|f_1\|^2 - A_4 \left(\|f\|^3 + \|f\|^4\right)$$

$$\geq A_5 \left(p^2 + \|f_1\|^2\right) - A_4 \left(\|f\|^3 + \|f\|^4\right)$$

$$= A_5 \|f\|^2 - A_4 \left(\|f\|^3 + \|f\|^4\right), \tag{6.164}$$

where $A_5 = \min(\alpha, \beta) > 0$ (β is defined in (6.150)) and $A_4 = A_3 \left(\frac{5c}{4} + \alpha\right) > 0$. Again, we note that A_4 and A_5 are independent of $f(x)$ and a.

Finally, if we let θ be a real number satisfying $0 < \theta < 1$ and form

$$\theta \times \text{Eqn}(6.154) + (1 - \theta) \times \text{Eqn}(6.164),$$

we get

$$(f, \mathcal{L}f) \geq \theta \|f\|_1^2 + [A_5(1 - \theta) - (1 + 2c)\theta] \|f\|^2$$

$$-A_4(1 - \theta) \left(\|f\|^3 + \|f\|^4\right).$$

And if we choose θ so that

$$A_5(1 - \theta) > (1 + 2c)\theta,$$

it follows that

$$(f, \mathcal{L}f) \geq A_1 \|f\|_1^2 - A_2 \left(\|f\|_1^3 + \|f\|_1^4\right), \tag{6.165}$$

where $A_1 = \theta > 0$ and $A_2 = A_4(1 - \theta) > 0$ and we have used $\|f\| \leq \|f\|_1$. This completes the proof. ∎

We are, however, unable to apply Lemma 6.8, as is, to establishing a lower bound for $(\varphi, \mathcal{L}_a\varphi)$ and hence $\Delta \mathcal{H}(\varphi)$ since $\varphi(x, t)$ does not, in general, satisfy the conditions of Lemma 6.8. The following result connects the assumptions of Lemma 6.8 to $\varphi(x, t)$.

Lemma 6.9 *Suppose $q(x, t)$ solves the KdV equation (6.84) subject to the initial condition $q(x, 0) = q_0(x)$ at least on $\Omega_\infty \times [0, T)$ where $0 < T \leq \infty$. Let $\varphi(x, t)$ be given by (6.137) and let $\widetilde{\varphi}(x, t)$ be defined by*

$$\widetilde{\varphi}(x, t) = \varphi(x, t) + b(t) q_s(x + a(t)), \tag{6.166}$$

where $a(t)$ and $b(t)$ are real valued functions defined at least for $t \in [0, T)$ chosen so that

$$\|q_s(x + a(t)) + \widetilde{\varphi}(x, t)\| = \|q_s\|, \tag{6.167}$$

$$\int_{\Omega_\infty} q_{s_x}(x + a(t)) \widetilde{\varphi}(x, t) \ dx = 0, \tag{6.168}$$

for all $t \in [0, T)$. Then there exist positive constants $\widetilde{A}_1$ and $\widetilde{A}_2$, which depend on $q_s(x)$ but are independent of $a(t)$, $b(t)$, $q(x, t)$ and $q_0(x)$ such that

$$\Delta \mathcal{H}(\varphi) \geq \Phi(\|\varphi\|_1) - \gamma(|b(t)|), \tag{6.169}$$

where

$$\Phi(z) = \widetilde{A}_1 z^2 - \widetilde{A}_2(z^3 + z^4), \tag{6.170}$$

$$\gamma(z) = \widetilde{A}_2(z^2 + z^4). \tag{6.171}$$

Proof. Our proof will assume the existence of $a(t)$ and $b(t)$ satisfying (6.167) and (6.168). In Lemma 6.10 we will establish the existence of the said $a(t)$ and $b(t)$. From (6.145) we have

$$\Delta \mathcal{H}(\varphi) \geq (\varphi, \mathcal{L}_a \varphi) - \frac{\|\varphi\|_1^3}{3\sqrt{2}}. \tag{6.172}$$

Using (6.166) we may write $(\varphi, \mathcal{L}_a \varphi)$ in the form

$$(\varphi, \mathcal{L}_a \varphi) = \int_{-\infty}^{\infty} \widetilde{\varphi} \mathcal{L}_a \widetilde{\varphi} \ dx - 2b(t) \int_{-\infty}^{\infty} \widetilde{\varphi} \mathcal{L}_a q_s(x + a(t)) \ dx$$

$$+ b^2(t) \int_{-\infty}^{\infty} q_s(x + a(t)) \mathcal{L}_a q_s(x + a(t)) \ dx. \tag{6.173}$$

We will estimate each integral in (6.173) individually. Since by assumption $\widetilde{\varphi}(x, t)$ satisfies the conditions of Lemma 6.8, there exist positive constants A_1 and A_2, independent of $\widetilde{\varphi}(x, t)$ and $a(t)$, such that

$$\int_{-\infty}^{\infty} \widetilde{\varphi} \mathcal{L}_a \widetilde{\varphi} \ dx \geq A_1 \|\widetilde{\varphi}\|_1^2 - A_2 \left(\|\widetilde{\varphi}\|_1^3 + \|\widetilde{\varphi}\|_1^4 \right). \tag{6.174}$$

For the second integral on the right hand side of (6.173) we have

$$2b(t) \int_{-\infty}^{\infty} \widetilde{\varphi} \mathcal{L}_a q_s(x + a(t)) \ dx \leq 2 |b| \, \|\widetilde{\varphi}\| \, \|\mathcal{L}_a q_s(x + a(t))\|$$

$$= 2 \left\| \mathcal{L}q_s \left(x \right) \right\| \left| b \right| \left\| \widetilde{\varphi} \right\| \le A_3 \left| b \right| \left\| \widetilde{\varphi} \right\|_1, \tag{6.175}$$

where

$$A_3 = 2 \left\| \mathcal{L}q_s \left(x \right) \right\| > 0,$$

which depends only on q_s.

For the third integral on the right hand side of (6.173) we have

$$b^2 \left(t \right) \int_{-\infty}^{\infty} q_s \left(x + a \left(t \right) \right) \mathcal{L}_a q_s \left(x + a \left(t \right) \right) \ dx$$

$$= b^2 \int_{-\infty}^{\infty} q_s \left(x \right) \mathcal{L}q_s \left(x \right) \ dx \le b^2 \left\| \mathcal{L}q_s \left(x \right) \right\| \left\| q_s \left(x \right) \right\| = A_4 \left| b \right|^2, \tag{6.176}$$

where

$$A_4 = \left\| \mathcal{L}q_s \left(x \right) \right\| \left\| q_s \left(x \right) \right\| > 0,$$

which depends only on q_s.

The estimates (6.174), (6.175) and (6.176) together with (6.172) imply

$$\Delta \mathcal{H} \left(\varphi \right) \ge A_1 \left\| \widetilde{\varphi} \right\|_1^2 - A_3 \left| b \right| \left\| \widetilde{\varphi} \right\|_1 - A_4 \left| b \right|^2$$

$$- \widehat{A}_2 \left(\left\| \widetilde{\varphi} \right\|_1^3 + \left\| \widetilde{\varphi} \right\|_1^4 \right), \tag{6.177}$$

where

$$\widehat{A}_2 = A_2 + \frac{1}{3\sqrt{2}} > 0.$$

To proceed further we need to estimate each individual term in (6.177) as follows. For $\left\| \widetilde{\varphi} \right\|_1$ we have

$$\left\| \widetilde{\varphi} \right\|_1 = \left\| \varphi \left(x, t \right) + b \left(t \right) q_s \left(x + a \left(t \right) \right) \right\|_1$$

$$\le \left\| \varphi \right\|_1 + \left| b \left(t \right) \right| \left\| q_s \left(x + a \left(t \right) \right) \right\|_1$$

$$= \left\| \varphi \right\|_1 + \left| b \left(t \right) \right| \left\| q_s \left(x \right) \right\|_1 = \left\| \varphi \right\|_1 + A_5 \left| b \right|,$$

where $A_5 = \left\| q_s \left(x \right) \right\|_1 > 0$. Similarly,

$$\left\| \widetilde{\varphi} \right\|_1 \ge \left\| \varphi \right\|_1 - A_5 \left| b \right|.$$

Thus,

$$A_1 \|\widetilde{\varphi}\|_1^2 - A_3 |b| \|\widetilde{\varphi}\|_1 \geq A_1 \left(\|\varphi\|_1 - A_5 |b| \right)^2 - A_3 |b| \left(\|\varphi\|_1 + A_5 |b| \right)$$

$$\geq A_1 \|\varphi\|_1^2 - A_6 |b| \|\varphi\|_1 - A_7 |b|^2 ,$$

where

$$A_6 = 2A_1 A_5 + A_3 > 0,$$

$$A_7 = A_3 A_5 > 0.$$

Using Young's inequality (see, e.g., Royden, 1968) in the form

$$|b| \|\varphi\|_1 \leq \frac{1}{2} \left(\theta \|\varphi\|_1^2 + \frac{|b|^2}{\theta} \right),$$

for any real number $\theta > 0$, implies

$$A_1 \|\widetilde{\varphi}\|_1^2 - A_3 |b| \|\widetilde{\varphi}\|_1 - A_4 |b|^2$$

$$\geq \left(A_1 - \frac{A_6 \theta}{2} \right) \|\varphi\|_1^2 - \left(A_7 + \frac{A_6}{2\theta} + A_4 \right) |b|^2 .$$

We can always choose θ sufficiently small and positive so that

$$\widetilde{A}_1 = A_1 - \frac{A_6 \theta}{2} > 0.$$

Hence we have positive constants $\widetilde{A}_1$ and

$$A_8 = A_7 + \frac{A_6}{2\theta} + A_4 > 0,$$

depending only on q_s, such that

$$A_1 \|\widetilde{\varphi}\|_1^2 - A_3 |b| \|\widetilde{\varphi}\|_1 - A_4 |b|^2 \geq \widetilde{A}_1 \|\varphi\|_1^2 - A_8 |b|^2 . \tag{6.178}$$

For $\widehat{A}_2 \|\widetilde{\varphi}\|_1^3$ we have

$$\widehat{A}_2 \|\widetilde{\varphi}\|_1^3 \leq \widehat{A}_2 \left(\|\varphi\|_1 + A_5 |b| \right)^3$$

$$= \widehat{A}_2 \left(\|\varphi\|_1^3 + 3A_5 \|\varphi\|_1^2 |b| + 3A_5^2 \|\varphi\|_1 |b|^2 + A_5^3 |b|^3 \right)$$

$$\leq \widehat{A}_2 \left(1 + 3A_5 + 2A_5^2\right) \|\varphi\|_1^3 + \widehat{A}_2 \left(A_5 + 2A_5^2 + A_5^3\right) |b|^3$$

$$\leq A_9 \left(\|\varphi\|_1^3 + |b|^2 + |b|^4\right), \tag{6.179}$$

where

$$A_9 = \widehat{A}_2 \max \left(1 + 3A_5 + 2A_5^2, \left(A_5 + 2A_5^2 + A_5^3\right)/2\right) > 0,$$

(which depends only on q_s) and we have repeatedly used Young's inequality in the forms

$$|f|\,|g| \leq \frac{2}{3} |f|^{\frac{3}{2}} + \frac{1}{3} |g|^3,$$

$$|b|^3 \leq \frac{1}{2} \left(|b|^2 + |b|^4\right).$$

Following a similar argument, there exists a positive constant A_{10}, depending only on q_s, such that

$$\widehat{A}_2 \|\widetilde{\varphi}\|_1^4 \leq A_{10} \left(\|\varphi\|_1^4 + |b|^4\right), \tag{6.180}$$

where we have repeatedly used Young's inequality in the form

$$|f|\,|g| \leq \frac{3}{4} |f|^{\frac{4}{3}} + \frac{1}{4} |g|^4.$$

Finally, substituting (6.178), (6.179) and (6.180) into (6.177) implies that

$$\Delta\mathcal{H}(\varphi) \geq \widetilde{A}_1 \|\varphi\|_1^2 - \widetilde{A}_2 \left(\|\varphi\|_1^3 + \|\varphi\|_1^4\right) - \widetilde{A}_2 \left(|b|^2 + |b|^4\right), \tag{6.181}$$

where

$$\widetilde{A}_2 = \max \left(A_8 + A_9, A_9 + A_{10}\right) > 0,$$

which depends only on q_s. The estimate (6.181) is exactly in the form (6.170), (6.171) and (6.172) as required. ∎

Provided we can construct functions $a(t)$ and $b(t)$ so that the function $\widetilde{\varphi}(x,t)$, defined by (6.166), can satisfy the conditions (6.167) and (6.168), Lemma 6.9 establishes the necessary lower bound for $\Delta\mathcal{H}(\varphi)$ from which we will able to establish stability. What we need to do now is prove that such $a(t)$ and $b(t)$ exist at least for some nontrivial interval of time.

Lemma 6.10 *Suppose that*

$$\|q_0(x) - q_s(x)\| \le \min\left[\frac{\|q_s\|}{15}, \frac{\|q_{s_x}\|^2}{2\|q_{s_{xx}}\|}\right], \tag{6.182}$$

then there exist, at least for a finite nontrivial interval of time, continuously differentiable functions $a(t)$ and $b(t)$ which satisfy (6.167) and (6.168) for $\tilde{\varphi}(x,t)$ defined by (6.166).

Proof. If (6.166) and (6.137) are substituted into (6.168), it follows that

$$\int_{-\infty}^{\infty} q_{s_x}(x + a(t)) q(x,t) \ dx = 0. \tag{6.183}$$

Thus, if we define

$$F(t,z) = \int_{-\infty}^{\infty} q_{s_x}(x + z) q(x,t) \ dx, \tag{6.184}$$

then $a(t)$ can be written as the solution of

$$F(t, a(t)) = 0. \tag{6.185}$$

Our approach here is to first establish that (6.185) has a solution at $t = 0$ and then derive an equivalent ordinary differential equation which will imply the existence of $a(t)$ for at least a finite interval of time. Let

$$\phi(z) = \int_{-\infty}^{\infty} [q_0(x) - q_s(x + z)]^2 \ dx \ge 0, \tag{6.186}$$

where we recall that $q_0(x) = q(x,0)$. Obviously, $\phi(z)$ is a continuously differentiable function for all $z \in \mathbb{R}$, i.e., $\phi \in C^1(\mathbb{R})$. Now it follows that

$$\lim_{|z| \to \infty} \phi(z) = \|q_0\|^2 + \|q_s\|^2 > \|q_s\|^2, \tag{6.187}$$

since both $q_0(x)$ and $q_s(x)$ are in $H^1(\mathbb{R})$. However, we also have, on account of (6.182)

$$\phi(0) = \|q_0 - q_s\|^2 \le \frac{\|q_s\|^2}{225} < \|q_s\|^2. \tag{6.188}$$

Since $\phi \in C^1(\mathbb{R})$, (6.186), (6.187) and (6.188) imply that the *minimum* of $\phi(z)$ exists and occurs at a finite value, say at $z = a_0$, which satisfies $\phi'(a_0) = 0$, i.e.,

$$0 = \int_{-\infty}^{\infty} [q_0(x) - q_s(x + a_0)] q_{s_x}(x + a_0) \ dx$$

$$= \int_{-\infty}^{\infty} q_0(x) \, q_{s_x}(x + a_0) \, dx = F(0, a(0)), \qquad (6.189)$$

where $a(0) = a_0$.

It is important to realize that the *a priori* estimate $\|q_0 - q_s\| < \|q_s\|$ is an essential aspect of the proof just given. This assumption does not jeopardize the possibility of proving Theorem 6.5 since if, for a given $\varepsilon > 0$, $\delta > 0$ can be found so that the theorem holds, then certainly there will exist δ values satisfying $0 < \delta < \|q_s\|$ or, for that matter,

$$0 < \delta < \min \left[\frac{\|q_s\|}{15}, \frac{\|q_{s_x}\|^2}{2 \, \|q_{s_{xx}}\|} \right].$$

Indeed, as pointed out by Bona (1975), an assumption such as $\|q_0 - q_s\| < \|q_s\|$ is necessary to ensure that the minimum of $\phi(z)$ occurs at a finite value. For example, without any restriction, one could assume that $q_0(x) < 0$ (e.g., $q_0 = -q_s$) and then the minimum of $\phi(z)$ occurs at infinity. The reason that we assume (6.182) specifically is that this choice, as we shown below, is *sufficient* to ensure the existence of $a(t)$ and $b(t)$ for a finite interval of time.

If (6.185) is differentiated with respect to t, one obtains the initial value problem

$$\frac{da}{dt} = \Lambda(t, a) \equiv \frac{\frac{\partial F(t,a)}{\partial t}}{\frac{\partial F(t,a)}{\partial z}}; \quad a(0) = a_0. \qquad (6.190)$$

Clearly, if $a(t)$ solves this problem, it also satisfies (6.185). We must determine conditions under which the right hand side of (6.190) implies the existence of a solution $a(t)$. We first note that the smoothness of $q(x, t)$ and $q_s(x)$ imply that both $\partial F(t, a)/\partial t$ and $\partial F(t, a)/\partial z$ are at least continuously differentiable.

Of concern is the possibility that there are values for which $\partial F(t, a)/\partial z = 0$. However, we can establish the following lower bound

$$\left| \frac{\partial F(t, a)}{\partial z} \right| = \left| \frac{\partial}{\partial z} \int_{-\infty}^{\infty} q_{s_x}(x + z) \, q(x, t) \, dx \right|$$

$$= \left| \int_{-\infty}^{\infty} q_{s_{xx}}(x + z) \left[q(x, t) - q_s(x + z) + q_s(x + z) \right] dx \right|$$

$$= \left| \int_{-\infty}^{\infty} q_{s_{xx}}(x + z) \left[q(x, t) - q_s(x + z) \right] - \left[q_{s_x}(x) \right]^2 \, dx \right|$$

$$\geq \|q_{s_x}\|^2 - \left| \int_{-\infty}^{\infty} q_{s_{xx}}(x + z) \left[q(x, t) - q_s(x + z) \right] dx \right|$$

$$\geq \|q_{s_x}\|^2 - \|q_{s_{xx}}\| \, \|q\,(x,t) - q_s\,(x+z)\|. \tag{6.191}$$

Thus,

$$\left| \frac{\partial F\,(t,a)}{\partial z} \right| > 0, \tag{6.192}$$

on the set

$$\Gamma = \left\{ (t,z) \mid \|q\,(x,t) - q_s\,(x+z)\| < \frac{\|q_{s_x}\|^2}{\|q_{s_{xx}}\|} \right\}. \tag{6.193}$$

And since

$$\|q_0\,(x) - q_s\,(x+a_0)\| = \sqrt{\phi\,(a_0)} \leq \sqrt{\phi\,(0)}$$

$$= \|q_0\,(x) - q_s\,(x)\| \leq \frac{\|q_{s_x}\|^2}{2\,\|q_{s_{xx}}\|} < \frac{\|q_{s_x}\|^2}{\|q_{s_{xx}}\|},$$

it follows that $(0,a_0) \in \Gamma$. Hence $\Lambda\,(t,a)$ is defined and continuously differentiable on Γ which contains within it $(0,a_0)$, and thus by the classical existence theory for ordinary differential equations, e.g., Murray & Miller (1976), there exists a solution to the initial value problem (6.190) for some nontrivial interval of time.

If we define T_0 to be the *largest* time (which could be infinite) for which a solution exists to (6.190), then we have established the existence of $a\,(t)$, in the interval $t \in [0, T_0)$, satisfying

$$0 = F\,(t, a\,(t)) = \int_{-\infty}^{\infty} q_{s_x}\,(x + a\,(t))\, q\,(x,t)\ dx$$

$$= \int_{-\infty}^{\infty} q_{s_x}\,(x + a\,(t))\,[q\,(x,t) - q_s\,(x + a\,(t))]\ dx$$

$$= \int_{-\infty}^{\infty} q_{s_\tau}\,(x + a\,(t))\,\varphi\,(x,t)\ dx, \tag{6.194}$$

where $\varphi\,(x,t)$ is defined by (6.137).

Equation (6.194) is equivalent to (6.168) provided $b\,(t)$ exists. If (6.137) and (6.166) are substituted into (6.167), it follows that

$$\|q\,(x,t) + b\,(t)\, q_s\,(x + a\,(t))\|^2 = \|q_s\|^2, \tag{6.195}$$

which can expanded to give the quadratic equation

$$b^2 + 2r(t)b + \frac{\|q\|^2 - \|q_s\|^2}{\|q_s\|^2} = 0, \tag{6.196}$$

where

$$r(t) = \frac{(q(x,t), q_s(x + a(t)))}{\|q_s\|^2}. \tag{6.197}$$

The solutions for $b(t)$ may be expressed in the form

$$b(t) = -r(t) \pm \left(r^2(t) - \frac{\|q\|^2 - \|q_s\|^2}{\|q_s\|^2}\right)^{\frac{1}{2}}. \tag{6.198}$$

The issue to be resolved here is whether or not $b(t)$ is a real-valued continuous function at least for some finite nontrivial interval of time. Let us define T_1 as

$$T_1 = \sup\{t \mid t \leq T_0 \text{ and } \|\varphi\| < \|q_s\|/2 \text{ on } [0,t)\}. \tag{6.199}$$

It follows that $T_1 > 0$ because of the continuity of φ and initially

$$\|\varphi\| \leq \|q_0(x) - q_s(x)\| \leq \|q_s\|/15 < \|q_s\|/2.$$

If (6.137) is used to eliminate $q(x,t)$ in (6.197) we get

$$r(t) = \frac{\|q_s\|^2 + (\varphi(x,t), q_s(x + a(t)))}{\|q_s\|^2}$$

$$\geq \frac{\|q_s\|^2 - |(\varphi(x,t), q_s(x + a(t)))|}{\|q_s\|^2}$$

$$\geq 1 - \frac{\|\varphi\|}{\|q_s\|} > \frac{1}{2}, \tag{6.200}$$

for $t \in [0, T_1)$.

On the other hand, exploiting the invariance of $\|q\|^2$, we have,

$$\frac{\|q\|^2 - \|q_s\|^2}{\|q_s\|^2} = \frac{\|q_0\|^2 - \|q_s\|^2}{\|q_s\|^2}$$

$$\leq \frac{\|q_0 - q_s\|(\|q_0\| + \|q_s\|)}{\|q_s\|^2}$$

$$\leq \frac{\|q_0 - q_s\| \left(\|q_0 - q_s\| + 2\|q_s\|\right)}{\|q_s\|^2}$$

$$< \frac{3\|q_0 - q_s\|}{\|q_s\|} < \frac{1}{5}, \tag{6.201}$$

where we have used the inequalities

$$\|q_0\| \leq \|q_0 - q_s\| + \|q_s\|,$$

$$\|q_0 - q_s\| \leq \|q_s\|/15 < \|q_s\|.$$

It follows, therefore, from (6.200) and (6.201) that

$$r^2(t) > \frac{\|q\|^2 - \|q_s\|^2}{\|q_s\|^2},$$

for $t \in [0, T_1)$ and thus $b(t)$ is a real-valued function (provided we take the + root in (6.198)). As well, since $a(t)$ has been shown to be continuous, so is $b(t)$.

Moreover, we can establish the following *a prior* estimate for $b(t)$ for $t \in [0, T_1)$. From (6.198), (6.200) and (6.201) it follows that

$$b(t) = \frac{\|q\|^2 - \|q_s\|^2}{\|q_s\|^2 \left[r(t) + \left(r^2(t) - \frac{\|q\|^2 - \|q_s\|^2}{\|q_s\|^2}\right)^{\frac{1}{2}}\right]},$$

so that

$$|b(t)| \leq \frac{\left|\|q\|^2 - \|q_s\|^2\right|}{r(t)\|q_s\|^2} \leq \frac{6\|q_0 - q_s\|}{\|q_s\|} \leq \frac{6\|q_0 - q_s\|_1}{\|q_s\|}. \tag{6.202}$$

Hence, we have shown that there exists $a(t)$ and $b(t)$, at least for $t \in [0, T_1)$, which satisfy (6.194) and (6.195). But these imply that (6.167) and (6.168) hold for $\tilde{\varphi}$ defined by (6.166), at least for $t \in [0, T_1)$ with $T_1 > 0$, since

$$0 = \int_{-\infty}^{\infty} q_{s_x}(x + a(t))\,\varphi(x,t)\,dx$$

$$= \int_{-\infty}^{\infty} q_{s_x}(x + a(t))\,[\tilde{\varphi}(x,t) - b(t)\,q_s(x + a(t))]\,dx,$$

$$= \int_{-\infty}^{\infty} q_{s_x}(x + a(t))\,\tilde{\varphi}(x,t)\,dx,$$

and

$$\|q_s\| = \|q(x,t) + b(t) q_s(x + a(t))\|$$

$$= \|q_s(x + a(t)) + \varphi(x,t) + b(t) q_s(x + a(t))\|$$

$$= \|q_s(x + a(t)) + \tilde{\varphi}(x,t)\|.$$

This completes the proof. ∎

Proof of Theorem 6.5.

We are now in a position to complete the proof of Theorem 6.5. Suppose we are given an $\varepsilon > 0$. We must find a $\delta_* > 0$ such that if $0 < \delta < \delta_*$ and $q_0(x) \in H^\infty(\Omega_\infty)$, $\|q_0(x) - q_s(x)\|_1 < \delta$ implies $d(q, q_s) < \varepsilon$ for all $t > 0$.

If we examine (6.170), we see that $\Phi(0) = \Phi'(0) = 0$ and $\Phi''(0) = 2\tilde{A}_1 > 0$. It follows that there exists a real number $\delta_1 > 0$ such that $\Phi(z)$ is strictly increasing for $z \in (0, \delta_1]$. Examining (6.171) we immediately see that $\gamma(z)$ is a strictly increasing continuous function for $z \in (0, \delta_1]$ and that $\gamma(0) = 0$.

In addition, we have $\Delta\mathcal{H} \to 0$ as $\|q_0(x) - q_s(x)\|_1 \to 0$, since if we introduce

$$q(x,t) = q_s(x) + \hat{q}(x,t),$$

into (6.135), we obtain

$$\Delta\mathcal{H} = \int_{\Omega_\infty} (\hat{q}_x)^2 + [c - q_s(x)]\hat{q}^2 - \frac{\hat{q}^3}{3} \, dx. \tag{6.203}$$

Here, $\hat{q}(x,t)$ may be thought of as the finite-amplitude perturbation to the soliton $q_s(x)$. Note that $\hat{q} \neq \varphi$ as defined by (6.137) unless $a = 0$ (which, in general, will not occur). If

$$q(x,0) - q_s(x) = \hat{q}(x,0) = \hat{q}_0(x) \in H^1(\Omega_\infty),$$

it follows, exploiting the invariance of $\Delta\mathcal{H}$, that

$$-2c\|q_0 - q_s\|_1^2 - \frac{\|q_0 - q_s\|_1^3}{3\sqrt{2}} \leq \Delta\mathcal{H}$$

$$\leq \max(1, c)\|q_0 - q_s\|_1^2 + \frac{\|q_0 - q_s\|_1^3}{3\sqrt{2}}.$$

Hence we conclude that

$$\lim_{\|q_0 - q_s\|_1 \to 0} \triangle\mathcal{H} = 0.$$

Thus $|\triangle\mathcal{H}|$ can be made as small as one wants if $\|q_0 - q_s\|_1$ is made sufficiently small. In addition, from (6.202) we see that $\gamma(|b(t)|)$ can be made small as we want if $\|q_0 - q_s\|_1$ is made sufficiently small.

Let us define

$$0 < \delta_2 = \min\left[\varepsilon, \delta_1, \frac{\|q_{s_s}\|^2}{2\|q_{s_{xx}}\|}, \frac{\|q_s\|}{15}\right].$$

Because we can make $|\triangle\mathcal{H}|$ and γ as small as we like if $\|q_0 - q_s\|_1$ is made sufficiently small, it follows that there exists a real number $\delta_* > 0$ such that if $0 < \delta < \delta_*$, then

$$\delta < \delta_2, \tag{6.204}$$

$$\triangle\mathcal{H} + \gamma\left(\frac{6\delta}{\|q_s\|}\right) < \Phi(\delta_2), \tag{6.205}$$

for $\|q_0 - q_s\|_1 < \delta$ where (6.202) has been used.

Moreover, since

$$\|q_0 - q_s\| \leq \|q_0 - q_s\|_1 < \delta,$$

and

$$\delta < \delta_2 \leq \min\left[\frac{\|q_{s_s}\|^2}{2\|q_{s_{xx}}\|}, \frac{\|q_s\|}{15}\right],$$

it follows that the conditions for Lemma 6.10 hold and thus there exist $a(t)$ and $b(t)$ which satisfy (6.167) and (6.168) for $\widetilde{\varphi}(x,t)$ defined by (6.166) for $t \in (0, T_1]$ where T_1 is defined by (6.199). Thus, by Lemma 6.9, we have

$$\triangle\mathcal{H} \geq \Phi(\|\varphi\|_1) - \gamma(|b(t)|)$$

$$\geq \Phi(\|\varphi\|_1) - \gamma\left(\frac{6\delta}{\|q_s\|}\right), \tag{6.206}$$

for $t \in (0, T_1]$ where (6.202) has been used.

The inequalities (6.205) and (6.206) together imply

$$\Phi(\|\varphi\|_1) \leq \Phi(\delta_2). \tag{6.207}$$

Since $\Phi(z)$ is an increasing function for $z \in (0, \delta_1]$ and $\|\varphi\|_1$ is, at least, a continuous function for, at least, $t \in (0, T_1]$, for which

$$\|\varphi(x, 0)\|_1 < \delta_2,$$

it follows from (6.207) that

$$\|\varphi(x, t)\|_1 \leq \delta_2 < \varepsilon, \tag{6.208}$$

and thus, on account of (6.138),

$$d(q(x, t), q_s(x)) < \varepsilon, \tag{6.209}$$

for $t \in (0, T_1]$. That is, we have established the stability of the *KdV* soliton for $t \in (0, T_1]$.

In point of fact, (6.208) implies that $T_1 = T_0 = \infty$ since if it were not so leads to a contradiction. Suppose, in the first case, that $T_1 < \infty$ and $T_1 < T_0$. From (6.208) we have

$$\|\varphi\|_1 \leq \delta_2 \leq \frac{\|q_s\|}{15} < \frac{\|q_s\|}{2}.$$

But this contradicts the definition of T_1 in (6.199). The only other finite possibility is that $T_1 = T_0 < \infty$. It follows from (6.208) that

$$\|\varphi\|_1 \leq \delta_2 \leq \frac{\|q_{s_s}\|^2}{2\|q_{s_{xx}}\|} < \frac{\|q_{s_s}\|^2}{\|q_{s_{xx}}\|}.$$

But then we have that

$$\lim_{t \to T_0} (t, a(t)) \in \Gamma,$$

so that the solution to (6.190) can be extended for some finite interval of time beyond T_0. But this contradicts the definition of T_0. There being no other finite choice we must conclude that $T_1 = T_0 = \infty$.

Hence we have proven that for $\varepsilon > 0$ there exists $\delta > 0$ such that $\|q_0 - q_s\|_1 < \delta$ implies that $d(q(x, t), q_s(x)) < \varepsilon$ for all $t > 0$. This completes the proof. ∎

6.7   Exercises

Exercise 6.1 *Show that the KdV equation corresponds to the first order necessary condition for an extremal to the Lagrangian*

$$\mathsf{L} = \int_\Omega \chi_t \chi_x + \frac{(\chi_x)^3}{3} - (\chi_{xx})^2 \ dx,$$

with the Clebsch variable transformation $q = \chi_x$.

Exercise 6.2 *Show that the Hamiltonian structure for the KdV equation given by (6.36) and (6.37) is invariant under arbitrary translations in x. In the case where $\Omega = \Omega_o$, it understood that the translations are modulo $2x_o$.*

Exercise 6.3 *Show directly that the momentum functional*

$$M(q) = -\frac{1}{2} \int_\Omega q^2 \, dx,$$

is an invariant with respect to time for the KdV equation.

Exercise 6.4 *Verify directly that the Casimir*

$$C(q) = \int_\Omega q \, dx,$$

is an invariant with respect to time for the KdV equation.

Exercise 6.5 *Show that (6.50) and (6.51) constitutes an alternate Hamiltonian formulation of the KdV equation (6.33).*

Exercise 6.6 *Show that the conserved functional associated with the invariance of the Hamiltonian structure (6.36) and (6.37) to arbitrary translations in time is the negative of the Hamiltonian (6.36).*

Exercise 6.7 *Determine the invariant associated with the invariance of the Hamiltonian structure (6.50) and (6.51) to arbitrary translations in space and time, respectively. Determine the Casimir functionals.*

Exercise 6.8 *Show that the KdV equation (6.33) is invariant under the Galilean transformation*

$$q(x,t) = q_\infty + \tilde{q}\left(\tilde{x}, \tilde{t}\right),$$

where $\tilde{x} = x - q_\infty t$ and $\tilde{t} = t$.

Exercise 6.9 *Show that*

$$\delta^2 \widehat{\mathcal{H}}(q_s) = \int_{\Omega_o} (\delta q_x)^2 + (c - q_s)(\delta q)^2 \, dx,$$

for the periodic solutions.

Exercise 6.10 *Show that $\psi_i(\zeta)$, for $i = 1$, 2 and 3, given by (6.92) and (6.95) are all proportional to $\operatorname{sech}^2(\zeta) \tanh(\zeta)$ for $\lambda = 0$ and ± 2.*

Exercise 6.11 *Verify that the Hamiltonian functional associated with the KdV equation (6.84) is given by (6.80) with J given by (6.37).*

References

1. Ablowitz, M. J. & Clarkson, P. A., 1991, *Solitons, Nonlinear Evolution Equations and Inverse Scattering*, London Mathematical Society Lecture Note Series **149**, Cambridge University Press.

2. Ablowitz, M. J. & Segur, H., 1981, *Solitons and the Inverse Scattering Transform*, SIAM Studies in Applied Mathematics, SIAM.

3. Abramowitz, M. & Stegun, I. A., 1972, *Handbook of Mathematical Functions (8th edition)*, Dover.

4. Andrews, D. G., 1984, On the existence of nonzonal flows satisfying sufficient conditions for stability. *Geophys. Astrophys. Fluid Dynamics* **28**, 243-256.

5. Adams, R. A., 1975, *Sobolev Spaces*, Academic.

6. Albert, J. P., Bona, J. L. & Henry, D. B., 1987, Sufficient conditions for stability of solitary-wave solutions of model equations for long waves. *Physica* **24D**, 343-366.

7. Arfken, G. B. & Weber, H. J., 1995, *Mathematical Methods for Physicists (Fourth Edition)*, Academic.

8. Arnol'd, V. I., 1965, Conditions for nonlinear stability of stationary plane curvilinear flows of an ideal fluid. *Sov. Math.* **6**, 773-777.

9. Arnol'd, V. I., 1969, On an *a priori* estimate in the theory of hydrodynamical stability. *Transl. Amer. Math. Soc.* **79**, 267-269.

10. Arnol'd, V. I., 1978, *Mathematical Methods of Classical Mechanics*, Graduate Texts in Mathematics, Springer-Verlag.

11. Atherton, R. W. & Homsy, G. M., 1975, On the existence and formulation of variational principles for nonlinear differential equations. *Stud. Appl. Math.* **LIV**, 31-60.

12. Benjamin, T. B., 1972, The stability of solitary waves. *Proc. R. Soc. Lond.* **A 328**, 153-183.

13. Benjamin, T. B., 1974, Lectures on nonlinear wave motion. In *Nonlinear Wave Motion* (ed. A. C. Newell), 1974, Lectures in Applied Mathematics **15**, pp. 3-47. American Mathematical Society.

14. Benjamin, T. B., 1984, Impulse, flow force and variational principles. *IMA J. appl. Math.* **32**, 3-68.

15. Benjamin, T. B., Bona, J. L. & Mahoney, J. J., 1972, Model equations for long waves in nonlinear dispersive systems. *Phil. Trans. R. Soc. Lond.* **A 272**, 47-78.

16. Bennett, A. F. & Kloeden, P. E., 1980, The simplified quasigeostrophic equations: existence and uniqueness of strong solutions. *Mathematika* **27**, 287-311.

17. Bennett, A. F. & Kloeden, P. E., 1981, The dissipative quasigeostrophic equations. *Mathematika* **28**, 265-285.

18. Bennett, A. F. & Kloeden, P. E., 1982, The periodic quasigeostrophic equations: existence and uniqueness of strong solutions. *Proc. R. Soc. Edinburgh* **91A**, 185-203.

19. Bennett, D. P., Brown, R. W., Stansfield, S. E., Stroughair, J. D. & Bona, J. L.,

264

1983, The stability of internal solitary waves. *Proc. Camb. Phil. Soc.* **94**, 351-379.

20. Bona, J. L., 1975, On the stability of solitary waves. *Proc. R. Soc. Lond.* **A 344**, 363-374.

21. Bona, J. L. & Smith, R., 1975, The initial-value problem for the Korteweg-de Vries equation. *Phil. Trans. R. Soc. Lond.* **A 278**, 555-601.

22. Bona, J. L., Souganidis, P. E. & Strauss, W. A., 1987, Stability and instability of solitary waves of Korteweg-de Vries type. *Proc. R. Soc. Lond.* **A 411**, 395-412.

23. Bona, J. L. & Soyeur, A., 1994, On the stability of solitary-wave solutions of model equations for long waves. *J. Nonlinear Sci.* **4**, 449-470.

24. Butchart, N., Haines, K. & Marshall, J. C., 1989, A theoretical and diagnostic study of solitary waves and atmosphere blocking. *J. Atmos. Sci.* **46**, 2063-2078.

25. Carnevale, G. F. & Shepherd, T. G., 1990, On the interpretation of Andrews' theorem. *Geophys. Astrophys. Fluid Dynamics* **51**, 1-17.

26. Carnevale, G. F., Vallis, G. K., Purini, R. & Briscolini, M., 1988, The role of initial conditions in flow stability with an application to modons. *Phys. Fluids* **31**, 2567-2572.

27. Charney, J. G., 1947, The dynamics of long waves in a baroclinic westerly current. *J. Meteorol.* **4**, 135-163.

28. Chern, S.-J. & Marsden, J. E., 1990, A note on symmetry and stability for fluid flows. *Geophys. Astrophys. Fluid Dynamics* **51**, 19-26.

29. Christiansen, J. P. & Zabusky, N. J., 1973, Instability, coalescence and fission of finite area vortex structures. *J. Fluid Mech.* **61**, 219-243.

30. Courant, R. & Hilbert, D., 1962, *Methods of Mathematical Physics (Vol. 2)*, Wiley-Interscience.

31. Drazin. P. G., 1977, On the stability of cnoidal waves. *Quart. J. Mech. Appl. Math.* **30**, 91-105.

32. Drazin, P. G. & Johnson, R. S., 1989, *Solitons: an introduction*, Cambridge University Press.

33. Drazin, P. G. & Reid, W. H., 1981, *Hydrodynamic Stability*, Cambridge University Press.

34. Dutton, J. A., 1974, The nonlinear quasi-geostrophic equation: existence and uniqueness of solutions on a bounded domain. *J. Atmos. Sci.* **31**, 422-433.

35. Ek, N. & Swaters, G. E., 1994, Geostrophic scatter diagrams and the application of quasi-geostrophic free-mode theory to a northeast Pacific blocking episode. *J. Atmos. Sci.* **51**, 563-581.

36. Fermi, E., Pasta, J. & Ulam, S. M., 1955, Studies in problems. *Tech. Rep.*, **LA-1940**, Los Alamos Sci. Lab.

37. Fjortoft, R., 1950, Application of integral theorems in deriving criteria of stability for laminar flows and for the baroclinic circular vortex. *Geofys. Pub. Oslo* **17**(6), 1-52.

38. Flierl, G. R., 1988, On the instability of geostrophic vortices. *J. Fluid Mech.* **197**, 349-388.

39. Gardner, C. S., 1971, The Korteweg-de Vries equations and generalizations IV. The Korteweg-de Vries equation as a Hamiltonian system. *J. Math. Phys.* **12**, 97-133.

40. Gear, J. A. & R. Grimshaw, 1983: A second order theory for solitary waves in shallow fluids. *Phys. Fluids* **26**, 14 - 29.

41. Gelfand, I. & Fomin, S., 1963, *Calculus of Variations*, Prentice-Hall.

42. Goldstein, H., 1980, *Classical Mechanics*, Addison-Wesley.

43. Grimshaw, R., 1979, Slowly-varying solitary waves. I. Korteweg-de Vries equation. *Proc. R. Soc. Lond. A* **368**, 359-375.

44. Hasegawa, A. & Mima, K., 1978, Pseudo-three-dimensional turbulence in magnetized nonuniform plasma. *Phys. Fluids* **21**(1), 87-92.

45. Holm, D. D., Marsden, J. E., Ratiu, T. & Weinstein, A., 1985, Nonlinear stability of fluid a plasma equilibria. *Phys. Rep.* **123**, 1-116.

46. Holmes, P., Lumley, J. L. & Berkooz, G., 1998, *Turbulence, Coherent Structures, Dynamical Systems and Symmetry*. Cambridge University Press.

47. Jeffery, A. & Kakutani T., 1970, Stability of the Burgers shock wave and the Korteweg-de Vries soliton. *Indiana Univ. Math. J.* **20**, 463-468.

48. Jeffery, A. & Kakutani T., 1972, Weak nonlinear dispersive waves: A discussion centered around the Korteweg-de Vries equation. *SIAM Rev.* **14**, 582-643.

49. Judovic, V. I., 1966, A two-dimensional problem of unsteady flow of an ideal incompressible fluid across a given domain. *Transl. Amer. Math. Soc.* **17**, 277-304.

50. Kato, T., 1967, On classical solutions of the two-dimensional non-stationary Euler equation. *Arch. Rational Mech. Anal.* **25**, 188-200.

51. Kaup, D. J. & Newell, A. C., 1978, Solitons as particles, oscillators, and in slowly changing media: A singular perturbation theory. *Proc. R. Soc. Lond. A* **361**, 413-446.

52. Kinsman, B., 1965, *Water Waves*, Prentice-Hall.

53. Kloeden, P. E., 1987, On the uniqueness of solitary Rossby waves. *J. Austral. Math. Soc. B* **28**, 476-485.

54. Kloosterziel, R. C. & Carnevale, G. F., 1992, Formal stability of circular vortices. *J. Fluid Mech.* **242**, 249-278.

55. Korteweg, D. J. & de Vries, G., 1895, On the change of form of long waves advancing in a rectangular canal, and on a new type of long stationary waves. *Phil. Mag.* **39**, 422-443.

56. Kundu, P. K., 1990, *Fluid Mechanics*, Academic.

57. Kuroda, Y., 1990, Symmetries and Casimir invariants for perfect fluid. *Fluid Dyn. Res.* **5**, 273-287.

58. Ladyzhenskaya, O. A., 1969, *The Mathematical Theory of Viscous Incompressible Flow*, Gordon and Breach.

59. Laedke, E. W. & Spatschek, K. H., 1986, Two-dimensional drift-vortices and their stability. *Phys. Fluids* **29**, 133-142.

60. Lamb, H., 1932, *Hydrodynamics*, 6th Edition, Cambridge University Press.

61. Larichev, V. & Reznik, G., 1976, Two-dimensional Rossby soliton: an exact solution. *Rep. U.S.S.R. Acad. Sci.* **231**(5), 1077-1079.

62. LeBlond, P. H. & Mysak, L. A., 1978, *Waves in the Ocean*, Elsevier.

63. Lighthill, M. J., 1962, *Fourier Series and Generalized Functions*, Cambridge University Press.

64. Lorenz, E. N., 1972, Barotropic instabilities of Rossby wave motion. *J. Atmos. Sci.* **29**, 258-264.

65. Makino, M., Kamimura, T. & Taniuti, T., 1981, Dynamics of two-dimensional solitary vortices in a low-beta plasma with convective motion. *J. Phys. Soc. Japan* **50**, 980-989.

66. Marchioro, C. & Pulvirenti, M., 1994, *Mathematical theory of incompressible nonviscous fluids*, Springer-Verlag.

67. Maxwell, J. C., 1885a, Manuscript on the steady motion of an incompressible fluid. In *The Scientific Letters and Papers of James Clerk Maxwell* (ed. P. M. Hartman), 1990, Vol. 1., pp. 295-299. Cambridge University Press.

68. Maxwell, J. C., 1885b, Letter to William Thomson. In *The Scientific Letters and Papers of James Clerk Maxwell* (ed. P. M. Hartman), 1990, Vol. 1., pp. 309-313. Cambridge University Press.

69. McWilliams, J. C., Flierl, G. R., Larichev, V. D. & Reznik, G., 1981, Numerical studies of barotropic modons. *Dyn. Atmos. Oceans* **5**, 219-238.

70. McWilliams, J. C., & Zabusky, N. J., 1982, Interactions of isolated vortices. I. Modons colliding with modons. *Geophys. Astrophys. Fluid Dynamics* **19**, 207-227.

71. Miura, R. M., Gardner, C. S. & M. D. Kruskal, 1968, Korteweg-de Vries equation and generalizations. II. Existence of conservation laws and constants of motion. *J. Math. Phys.* **9**, 1204-1209.

72. Morrison, P. J., 1982, Poisson brackets for fluids and plasmas. In *Mathematical methods in hydrodynamics and integrability in dynamical systems* (ed. M. Tabor & Y. M. Treve), A. I. P. Conference Proceedings **88**, pp. 13-46.

73. Morrison, P. J. & Greene, J. M., 1980, Noncanonical Hamiltonian density formulation of hydrodynamics and ideal magnetohydrodynamics. *Phys. Rev. Lett.* **45**, 790-794.

74. Morse, P. & Feshbach, H., 1953, *Methods of Theoretical Physics*, McGraw-Hill.

75. Murray, F. J. & Miller, K. S., 1976, *Existence Theorems for Ordinary Differential Equations*, Krieger.

76. Navier, M., 1827, Mémoire sur les lois du mouvement des fluides. *Mém. de l'Acad. d. Sci.* **6**, 389-416.

77. Newell, A. C., 1985, *Solitons in Mathematics and Physics*, CBMS-NSF Regional Conference Series in Applied Mathematics **48**, SIAM.

78. Nycander, J., 1992, Refutation of stability proofs for dipole vortices. *Phys. Fluids* A **4**, 467-476.

79. Olver, P. J., 1982, A nonlinear Hamiltonian structure for the Euler equations. *J.*

appl. math. Analysis **89**, 233-250.

80. Pedlosky, J., 1987, *Geophysical Fluid Dynamics* (2nd Edition), Springer-Verlag.

81. Petviashvili, V. I., 1983, Cyclones and anticyclones in zonal flow. In *Nonlinear and Turbulent Processes in Physics* (ed. R. Z. Sagdev), Vol 1., pp. 979-987.

82. Poisson, S. D., 1831, Mémoire sur les équations générales de l'équilibre et du mouvement des corps solides élastiques et des fluides. *Jour. de l'Ecole polytechn.* **13**, 139-186.

83. Pierini, S., 1985, On the stability of equivalent modons. *Dyn. Atmos. Oceans* **9**, 273-280.

84. Rayleigh, Lord, 1880, On the stability, or instability, of certain fluid motions. *Proc. London Math. Soc.* **11**, 57-70.

85. Read, P. L., Rhines, P. B. & White A. A, 1986, Geostrophic scatter diagrams and potential vorticity dynamics. *J. Atmos. Sci.* **43**, 3226-3240.

86. Ripa, P., 1987, On the stability of elliptical vortex solutions of the shallow water equations. *J. Fluid Mech.* **183**, 343-346. *Phys. Fluids* **A 4**, 460-463.

87. Ripa, P., 1992, Comments on a paper by Sakuma and Ghil. *Phys. Fluids* **A 4**, 460-463.

88. Robinson, S. K., 1991, Coherent motions in the turbulent boundary layer. *Ann. Rev. Fluid Mech.* **23**, 601-639.

89. Royden, H. L., 1968, *Real Analysis (2nd Edition)*, Macmillan.

90. Rossby, C. G., 1939, Relation between variations in the intensity of the zonal circulations of the atmosphere and the displacements of the semi-permanent centers of action. *J. Mar. Res.* **2**, 38-55.

91. Sakuma, H. & Ghil, M., 1990, Stability of stationary barotropic modons for small-amplitude perturbations. *Phys. Fluids* **A 3**, 408-414.

92. Sakuma, H. & Ghil, M., 1990, Stability of propagating modons by Lyapunov's direct method. *J. Fluid Mech.* **211**, 393-416.

93. Salmon, R., 1988, Hamiltonian fluid mechanics. *Ann. Rev. Fluid Mech.* **20**, 225-256.

94. Scinocca, J. F. & Shepherd, T. G., 1992, Nonlinear wave-activity conservation laws and Hamiltonian structure for the two-dimensional inelastic equations. *J. Atmos. Sci.* **49**, 5-27.

95. Seliger, R. L. & Whitham, G. R., 1968, Variational principles in continuum mechanics. *Proc. R. Soc. Lond.* **A 305**, 1-25.

96. Shepherd, T. G., 1990, Symmetries, conservation laws, and Hamiltonian structure in geophysical fluid dynamics. *Adv. Geophys.* **32**, 287-338.

97. Stern, M. E., 1975, Minimal properties of planetary eddies. *J. Mar. Res.* **33**, 1-13.

98. Stokes, G. G., 1845, On the theories of internal friction of fluids in motion. *Trans. Cambr. Phil. Soc.* **8**, 287-305.

99. Swaters, G. E., 1986, Stability conditions and *a priori* estimates for equivalent barotropic modons. *Phys. Fluids* **29**, 1419-1422.

100. Swaters, G. E., 1989, A perturbation theory for the solitary-drift-vortex solutions of the Hasegawa-Mima equation. *J. Plasma Physics* **41**, 523-539.

101. Temam, R., 1983, *Navier-Stokes Equations and Nonlinear Functional Analysis*, CBMS-NSF Regional Conference Series in Applied Mathematics **41**, SIAM.

102. de St. Venant, B., 1843, Note à joindre un mémoire sur la dynamique des fluides. *Comptes-Rendus* **17**, 1240-1244.

103. Vladimirov, V. A., 1986, On nonlinear stability of incompressible fluid flows. *Arch. Mech. Stos.* **38**(5/6), 689-696.

104. Virasoro, M. A., 1981, Variational principle of two-dimensional incompressible hydrodynamics and quasi-geostrophic flows. *Phys. Rev. Lett.* **47**, 1181-1183.

105. Wan, Y. H. & M. Pulvirenti, 1985, Nonlinear stability of circular vortex patches. *Commun. Math. Phys.* **99**, 435-450.

106. Wan, Y. H., 1986, The stability of rotating vortex patches. *Commun. Math. Phys.* **107**, 1-20.

107. Weinstein, A., 1983, Hamiltonian structure for drift waves and geostrophic flow. *Phys. Fluids.* **26**, 388-390.

108. Weinstein, M. I., 1986, Lyapunov stability of ground states of nonlinear dispersive evolution equations. *Commun. Pure Appl. Math.* **39**, 51-68.

109. Weiss, J., 1991, The dynamics of enstrophy transfer in two-dimensional hydrodynamics. *Physica D* **48**, 273-294.

110. Wolibner, W., 1933, Un theoreme sur l'existence du movement plan d'un fluide parfait, homogene, incompressible, pendant un temps infiniment longue. *Math Z.* **37**, 698-726.

111. Zabusky, N. J. & Kruskal, M. D., 1965, Interactions of 'solitons' in a collisionless plasma and the recurrence of initial states. *Phys. Rev. Lett.* **15**, 240-243.

112. Zakharov, V. E. & Faddeev, L. D., 1971, The Korteweg-deVries equation: a completely integrable Hamiltonian system. *Funct. Anal. Appl.* **5**, 280-287.

113. Zauderer, E., 1983, *Partial Differential Equations of Applied Mathematics*, Wiley-Interscience.

Index

Milton Keynes UK
Ingram Content Group UK Ltd.
UKHW040445071024
449327UK00020B/1004